MATHEMATICS APPLIED TO ENGINEERING IN ACTION

Advanced Theories, Methods, and Models

MATHEMATICS APPLIED TO ENGINEERING IN ACTION

Advanced Theories, Methods, and Models

Edited by
Nazmul Islam, PhD
Satya Bir Singh, PhD
Prabhat Ranjan, PhD
A. K. Haghi, PhD

APPLE
ACADEMIC
PRESS

First edition published 2021

Apple Academic Press Inc.
1265 Goldenrod Circle, NE,
Palm Bay, FL 32905 USA

4164 Lakeshore Road, Burlington,
ON, L7L 1A4 Canada

CRC Press
6000 Broken Sound Parkway NW,
Suite 300, Boca Raton, FL 33487-2742 USA

2 Park Square, Milton Park,
Abingdon, Oxon, OX14 4RN UK

First issued in paperback 2021

© 2021 Apple Academic Press, Inc.

Apple Academic Press exclusively co-publishes with CRC Press, an imprint of Taylor & Francis Group, LLC

Library and Archives Canada Cataloguing in Publication

Title: Mathematics applied to engineering in action : advanced theories, methods, and models / edited by Nazmul Islam, PhD, Satya Bir Singh, PhD, Prabhat Ranjan, PhD, A.K. Haghi, PhD.
Names: Islam, Nazmul, editor. | Singh, Satya Bir, editor. | Ranjan, Prabhat, (Mechatronics professor), editor. | Haghi, A. K., editor.
Description: Includes bibliographical references and index.
Identifiers: Canadiana (print) 20200331426 | Canadiana (ebook) 20200331493 | ISBN 9781771889223 (hardcover) | ISBN 9781003055174 (ebook)
Subjects: LCSH: Engineering mathematics.
Classification: LCC TA330 .M38 2021 | DDC 620.001/51—dc23

Library of Congress Cataloging-in-Publication Data

Names: Islam, Nazmul, editor. | Singh, Satya Bir, editor. | Ranjan, Prabhat, editor. | Haghi, A. K., editor.
Title: Mathematics applied to engineering in action : advanced theories, methods, and models / edited by Nazmul Islam, PhD, Satya Bir Singh, PhD, Prabhat Ranjan, PhD, A. K. Haghi, PhD.
Description: First edition. | Palm Bay, FL : Apple Academic Press, Inc. ; Boca Raton, FL : CRC Press, 2021. | Includes bibliographical references and index. | Summary: "Mathematics Applied to Engineering in Action : Advanced Theories, Methods, and Models focuses on material relevant to solving the kinds of mathematical problems regularly confronted by engineers. This new volume explains how an engineer should properly define the physical and mathematical problem statements, choose the computational approach, and solve the problem by a proven reliable approach. It presents the theoretical background necessary for solving problems, including definitions, rules, formulas, and theorems on the particular theme. The book aims to apply advanced mathematics using real-world problems to illustrate mathematical ideas. This approach emphasizes the relevance of mathematics to engineering problems, helps to motivate the reader, and gives examples of mathematical concepts in a context familiar to the research students. The volume is intended for professors and instructors, scientific researchers, students, and industry professionals. It will help readers to choose the most appropriate mathematical modeling method to solve engineering problems"-- Provided by publisher.
Identifiers: LCCN 2020040919 (print) | LCCN 2020040920 (ebook) | ISBN 9781771889223 (hardcover) | ISBN 9781003055174 (ebook)
Subjects: LCSH: Engineering mathematics.
Classification: LCC TA330 .M3255 2021 (print) | LCC TA330 (ebook) | DDC 620.001/51--dc23
LC record available at https://lccn.loc.gov/2020040919
LC ebook record available at https://lccn.loc.gov/2020040920

ISBN: 978-1-77188-922-3 (hbk)
ISBN: 978-1-77463-775-3 (pbk)
ISBN: 978-1-00305-517-4 (ebk)

About the Editors

Nazmul Islam, PhD
Professor, Ramgarh Engineering College, Ramgarh, Jharkhand, India

Nazmul Islam, PhD, is now working as Professor at Ramgarh Engineering College, Jharkhand, India. He has published more than 60 research papers in several prestigious peer-reviewed journals and has written many book chapters and research books. In addition, he is the editor-in-chief of *The SciTech, Journal of Science* and *Technology, International Journal of Engineering Sciences*, and the *Signpost Open Access Journal of Theoretical Sciences*. He also serves as a member on the editorial boards of several journals. Dr. Islam's research interests are in theoretical chemistry, particularly quantum chemistry, conceptual density functional theory (CDFT), periodicity, SAR, QSAR/QSPR study, drug design, HMO theory, biological function of chemical compounds, quantum biology, nanochemistry, and more.

Satya Bir Singh, PhD
Professor, Department of Mathematics, Punjabi University, Patiala, India

Satya Bir Singh, PhD, is a Professor of Mathematics at Punjabi University Patiala in India. Prior to this, he worked as an Assistant Professor in Mathematics at the Thapar Institute of Engineering and Technology, Patiala, India. He has published about 125 research papers in journals of national and international repute and has given invited talks at various conferences and workshops. He has also organized several national and international conferences. He has been a coordinator and principal investigator of several schemes funded by the Department of Science and Technology, Government of India, New Delhi; the University Grants Commission, Government of India, New Delhi; and the All India Council for Technical Education, Government of India, New Delhi. He has 21 years of teaching and research experience. His areas of interest include mechanics of composite materials, optimization techniques, and numerical analysis. He is a life member of various learned bodies.

Prabhat Ranjan, PhD
Assistant Professor, Department of Mechatronics Engineering at Manipal University Jaipur, India

Prabhat Ranjan, PhD, is an Assistant Professor in the Department of Mechatronics Engineering at Manipal University Jaipur. He is the author of Basic Electronics and editor of *Computational Chemistry Methodology in Structural Biology and Materials Sciences*. Dr. Ranjan has published more than 10 research papers in peer-reviewed journals of high repute and dozens of book chapters in high-end research edited books. He has received prestigious the President Award of Manipal University Jaipur, India; a Material Design Scholarship from Imperial College of London, UK; a DAAD Fellowship; and the CFCAM-France Award. Dr. Ranjan has received several grants and also participated in national and international conferences and summer schools. He holds a bachelor of engineering in electronics and communication and a master of technology in instrumentation control system engineering from the Manipal Academy of Higher Education, Manipal, India, as well as a PhD in engineering from Manipal University Jaipur, India.

A. K. Haghi, PhD
Professor Emeritus of Engineering Sciences, Former Editor-in-Chief, International Journal of Chemoinformatics and Chemical Engineering and Polymers Research Journal; Member, Canadian Research and Development Center of Sciences and Culture

A. K. Haghi, PhD, is the author and editor of over 200 books, as well as over 1000 published papers in various journals and conference proceedings. Dr. Haghi has received several grants, consulted for a number of major corporations, and is a frequent speaker to national and international audiences. Since 1983, he served as professor at several universities. He is the former Editor-in-Chief of the *International Journal of Chemoinformatics and Chemical Engineering* and *Polymers Research Journal* and is on the editorial boards of many international journals. He is also a member of the Canadian Research and Development Center of Sciences and Cultures.

Contents

Contributors .... *ix*

Abbreviations .... *xiii*

Preface .... *xvii*

1. **Quantum Information Perspective on Chemical Reactivity** 1
   Roman F. Nalewajski

2. **A Computational Modeling of the Structure, Frontier Molecular Orbital (FMO) Analysis, and Global and Local Reactive Descriptors of a Phytochemical 'Coumestrol'** 41
   P. Vinduja, Vijisha K. Rajan, Swathi Krishna, and K. Muraleedharan

3. **Theoretical Analysis of CuTiS₂ and CuTiSe₂ Invoking Density Functional Theory-Based Descriptors** 61
   Prabhat Ranjan, Pancham Kumar, and Tanmoy Chakraborty

4. **Synergistic Effect of E. crassipes Biomass/Chitosan for As(III) Remediation from Water** 71
   Pankaj Gogoi, Pakiza Begum, Kaustubh Rakshit, and Tarun K. Maji

5. **Computational Investigations on Metal Oxide Clusters and Graphene-Based Nanomaterials for Heterogeneous Catalysis** 93
   Neetu Goel, Navjot Kaur, Mohd Riyaz, and Sarita Yadav

6. **Effect of Dye Concentration on Series Resistance of Thionin Dye-Based Organic Diode** 119
   Pallab Kumar Das, Swapan Bhunia, Sarmistha Basu, and N. B. Manik

7. **On Gower's Inverse Matrix** 133
   J. López-Bonilla, R. López-Vázquez, and S. Vidal-Beltrán

8. **Applications of Noether's Theorem** 141
   J. Yaljá Montiel-Pérez, J. López-Bonilla, R. López-Vázquez, and S. Vidal-Beltrán

9. Mathematical Modeling of Elastic-Plastic Transitional Stresses in
    Human Femur and Tibia Bones Exhibiting Orthotropic
    Macro Structural Symmetry.. 155
    Shivdev Shahi and Satya Bir Singh, PhD

10. Lorentz Transformations, Dirac Matrices, and 3-Rotations
    via Quaternions.. 171
    J. Morales, G. Ovando, J. López-Bonilla, and R. López-Vázquez

11. Multiscale Approach Towards the Modeling and Simulation
    of Carbon Nanotubes Networks 179
    Manas Roy and Mitali Saha

12. When Perovskites Memorize.. 193
    P. Sidharth, M. Prateek, and P. Predeep

Index.. 213

Contributors

Sarmistha Basu
Lecturer, Department of Electronics, Behala College, Behala, Kolkata – 700060, West Bengal, India

Pakiza Begum
Department of Chemistry, Indian Institute of Technology Guwahati, North Guwahati, Kamrup – 781039, Assam, India, E-mail: pakiza@iitg.ac.in

Swapan Bhunia
Assistant Professor, Department of Physics, Ramakrishna Residential College (Autonomous), Narendrapur, Kolkata – 700103, West Bengal, India

Tanmoy Chakraborty
Department of Chemistry, School of Engineering, Presidency University, Bengaluru – 560064, India, E-mail: tanmoychem@gmail.com

Pallab Kumar Das
Assistant Professor, Department of Electronics, Behala College, Behala, Kolkata – 700060, West Bengal, India

Neetu Goel
Theoretical and Computational Chemistry Group, Department of Chemistry and Center of Advanced Studies in Chemistry, Panjab University, Chandigarh – 160014, India, E-mail: neetugoel@pu.ac.in

Pankaj Gogoi
Department of Chemistry, Sipajhar College, Darrang – 784145, Assam, India, E-mail: pankaj5@yahoo.com

Navjot Kau
Theoretical and Computational Chemistry Group, Department of Chemistry and Center of Advanced Studies in Chemistry, Panjab University, Chandigarh – 160014, India

Swathi Krishna
Department of Chemistry, University of Calicut, Malappuram –673635, Kerala, India

Pancham Kumar
School of Electrical Skills, Bhartiya Skill Development University, Jaipur – 302042, India

J. López-Bonilla
Superior School of Electrical and Mechanical Engineering, National Polytechnic Institute, Building 4, Lindavista CP 07738, Mexico City, Mexico, E-mail: jlopezb@ipn.mx

R. López-Vázquez
Superior School of Electrical and Mechanical Engineering, National Polytechnic Institute, Building 4, Lindavista CP 07738, Mexico City, Mexico

Tarun K. Maji
Department of Chemical Sciences, Tezpur University, Napaam, Tezpur, Sonitpur – 784028, Assam, India, Tel.: +91 3712 267007, ext 5053, Fax: +91 3712 267005, E-mail: tkm@tezu.ernet.in

N. B. Manik
Condensed Matter Physics Research Center, Department of Physics, Jadavpur University, Kolkata – 700032, India, Tel.: +91 9831209230, E-mail: nb_manik@yahoo.co.in

J. Yaljá Montiel-Pérez
Computer Research Center, National Polytechnic Institute, CP 07738, Mexico City, Mexico

J. Morales
Area of Physics, Autonomous Metropolitan University, Azcapotzalco, St. Paul Ave. 180, CP 02200, Mexico City, Mexico

K. Muraleedharan
Department of Chemistry, University of Calicut, Malappuram – 673635, Kerala, India

Roman F. Nalewajski
Professor Emeritus, Department of Theoretical Chemistry, Jagiellonian University, Gronostajowa 2, 30–387 Cracow, Poland, E-mail: nalewajs@chemia.uj.edu.pl

G. Ovando
Area of Physics, Autonomous Metropolitan University, Azcapotzalco, St. Paul Ave. 180, CP 02200, Mexico City, Mexico

M. Prateek
Laboratory for Molecular Electronics and Photonics (LAMP, Department of Physics, National Institute of Technology, Calicut, Kerala, India

P. Predeep
Laboratory for Molecular Electronics and Photonics (LAMP, Department of Physics, National Institute of Technology, Calicut, Kerala, India, E-mail: predeep@nitc.ac.in

Vijisha K. Rajan
Department of Chemistry, University of Calicut, Malappuram – 673635, Kerala, India

Kaustubh Rakshit
Center for the Environment, Indian Institute of Technology Guwahati, North Guwahati, Kamrup – 781039, Assam, India, E-mail: kaustubhrakshit@gmail.com

Prabhat Ranjan
Department of Mechatronics Engineering, Manipal University Jaipur, Dehmi Kalan, Jaipur – 303007, India, E-mail: prabhat23887@gmail.com

Mohd Riyaz
Theoretical and Computational Chemistry Group, Department of Chemistry and Center of Advanced Studies in Chemistry, Panjab University, Chandigarh – 160014, India

Manas Roy
Department of Chemistry, National Institute of Technology, Agartala – 799046, Tripura, India

Mitali Saha
Associate Professor, Department of Chemistry, National Institute of Technology, Agartala – 799046, Tripura, India, Tel.: +91-8974006400, E-mail: mitalichem71@gmail.com

Shivdev Shahi
Department of Mathematics, Punjabi University Patiala, Punjab – 147002, India, E-mail: shivdevshahi93@gmail.com

P. Sidharth
Laboratory for Molecular Electronics and Photonics (LAMP, Department of Physics, National Institute of Technology, Calicut, Kerala, India

Satya Bir Singh, PhD
Department of Mathematics, Punjabi University Patiala, Punjab – 147002, India, E-mail: sbsingh69@yahoo.com

S. Vidal-Beltrán
Superior School of Electrical and Mechanical Engineering, National Polytechnic Institute, Building 4, Lindavista CP 07738, Mexico City, Mexico

P. Vinduja
Department of Chemistry, University of Calicut, Malappuram – 673635, Kerala, India

Sarita Yadav
Theoretical and Computational Chemistry Group, Department of Chemistry and Center of Advanced Studies in Chemistry, Panjab University, Chandigarh – 160014, India

Abbreviations

AIM	atoms-in-molecules
Al	aluminum
AO	atomic orbital
CB	conduction band
CG	contragradience
CMOS	complementary metal-oxide-semiconductor
CNTs	carbon nanotubes
CSA	charge sensitivity analysis
CSIF	Central Sophisticated Instrumentation Facility
CT	charge-transfer
CT	chitosan
CTCB	communication theory of the chemical bond
CTRB	chitosan root powder physical blend
CTRP	chitosan root powder composite
DFT	density functional theory
DNP	double numerical plus polarization
DVG	divacant-graphene
EA	electron affinity
EC	ethylene carbonate
Ef	Fermi energy
ELF	electron localization function
EO	equidensity orbitals
ER	Eley-Rideal
FF	Fukui function
FMO	frontier molecular orbital
FTIR	Fourier transformed infrared spectrophotometer
FTO	fluorine-doped indium tin oxide
GGA	generalized gradient approximation
GO	graphene oxide
GS	ground-state
HOMO	highest occupied molecular orbital
HRS	high resistance state
HSAB	hard soft acid and bases

HZM	Harriman-Zumbach-Maschke
IP	ionization potential
IR	infrared
IT	information theory
ITO	indium tin oxide
LEDs	light-emitting diodes
LH	Langmuir-Hinshelwood
LiClO4	lithium perchlorate
LRS	low resistance state
LSDA	local spin density approximation
LUMO	lowest unoccupied molecular orbital
MA	methylammonium
MAPbI3	methylammonium lead triiodide
MD	molecular dynamic
MM	molecular mechanics
MOFs	metal-organic frameworks
MWCNTs	multi-walled CNTs
NaAsO2	sodium metaarsenite
NEGF	non-equilibrium greens function
NMR	nuclear magnetic resonance
NO_2	nitrogen oxides
NRR	nitrogen reduction reaction
OBS	Ortmann, Bechstedt, and Schmidt
OCT	orbital communication theory
OER	oxygen evolution reaction
ORR	oxygen reduction reaction
PC	propylene carbonate
PEO	polyethylene oxide
PES	potential energy surface
PVA	polyvinyl alcohol
QIT	quantum information theory
QM	quantum mechanical
RC	reaction coordinate
RRAM	resistive random access memory
Rs	resistance
SCLC	space charge limited conduction
SE	Schrödinger equation
SO_2	sulfur dioxide

SWCNTs	single-walled CNTs
TCAD	technology computer-aided design
TDDFT	time-dependent density functional theory
TE	thermionic emission
TEM	transmission electron microscopic
TFL	trap-filled limit
TGA	thermogravimetric analysis
TS	transition-state
VB	valence band
vdW	van der Waals
VEA	vertical electron affinity
VIP	vertical ionization potential
VTFL	trap-filled limited voltage
WHRP	water hyacinth root powder
WORM	write once read many

Preface

The book particularly focuses on material relevant to solving the kinds of mathematical problems regularly confronted by engineers. This new volume explains how an engineer should properly define the physical and mathematical problem statements, choose the computational approach, and solve the problem by a proven reliable approach. It presents the theoretical background necessary for solving problems, including definitions, rules, formulas, and theorems on a particular theme.

The book aims to apply advanced mathematics using real-world problems to illustrate mathematical ideas. This approach emphasizes the relevance of mathematics to engineering problems, helps to motivate the reader, and gives examples of mathematical concepts in a context familiar to the research students.

It is intended for professors and instructors, scientific researchers, students, and industry professionals, the book will help readers to choose the most appropriate mathematical modeling method to solve engineering problems.

In the first chapter of this book, the need for quantum information measures combining the probability and current contributions is emphasized and continuity relations are summarized. Resultant gradient information accounts for classical (probability) and nonclassical (phase) contributions and reflects the kinetic energy content of the molecular electronic state. The phase-information description enables one to distinguish the hypothetical states of the mutually bonded (entangled) and nonbonded (disentangled) reactants in donor-acceptor systems. This generalized information-theoretic perspective is applied to chemical reactivity phenomena within the *grand*-ensemble description of thermodynamic conditions in open molecular systems. The physical equivalence of variational principles for the system electronic energy and the resultant information is demonstrated. The virial theorem decomposition of energy profiles is used to index the Hammond rule of reactivity theory and information changes in chemical reactions are addressed and charge transfer between reactants is explored. The *frontier*-electron approach to molecular interactions is used to rationalize the hard (soft) acid and bases (HSAB) principle of structural chemistry, and the rule implications for the *intra-* and *inter-*reactant communications are commented upon.

In Chapter 2, it is intended to show that coumestrol is an estrogenic isoflavonoid which belongs to the class of Phytochemicals known as coumestans, with interesting therapeutic applications such as antioxidant and anticancer properties. Coumestrol is widely spread in leguminous plants, alfalfa, ladino clover, strawberry, soya bean, sprouts, pea silage, and beans. The structure of coumestrol is closely resembles to the (*E*)-4, 4'-dihydroxystilbene derivatives, which have promising pharmacological activity, and are responsible for the estrogenic activity of coumestrol. Chapter 2, deals with the computational investigation of the structural analysis of coumestrol, by using M06 as a level of theory and 6–31++G (d, p) as a basis set in Gaussian 09 software package. The stable conformer of coumestrol has been identified through the potential energy scan (PES) and the lowest energy conformer is selected for further investigation. Before going to the deep knowledge of bioactivities shown by the title compound, one must have in-depth knowledge about the molecular structure of the compound. The most robust computational tool, density functional theory (DFT) has been enhanced to a great extent, especially for the structural analysis of organic compounds, a computational exploration into the structural analysis of the title compound is particularly relevant. Detailed structural analysis has been done using H^1 and C^{13} NMR as well as UV-visible spectroscopy. Moreover, the reports showing the detailed structural characterization of coumestrol is found to be rare and the reported papers are mostly focused on its bioactivities. In this scenario, the present work aims to explore a detailed computational approach towards the structural characterization of coumestrol. The work clearly explains the molecular structure, spectral characterizations, and frontier molecular orbital analysis. In addition, the global reactive and Fukui indices of the title compound have also been explained. The work can be extended to detailed bioactivity analysis, molecular docking analysis, QSAR/QSPR studies, etc.

In Chapter 3, we have systematically investigated the physicochemical properties of semiconducting materials $CuTiS_2$ and $CuTiSe_2$ in terms of DFT-based descriptors. The study of $CuTiS_2$ and $CuTiSe_2$ materials are of importance due to its potential applications in solar cells, light-emitting diodes (LEDs), and nonlinear optical devices. DFT is one of the most successful techniques in computational material science and engineering to compute the stability, structure, and electronic, optical, and magnetic properties of materials. Geometry optimization is done with exchange-correlation functional local spin density approximation (LSDA) and basis set Lanl2DZ. The DFT based descriptors namely: highest occupied molecular orbital (HOMO)-lowest unoccupied molecular orbital (LUMO), electronegativity, hardness, softness, electrophilicity index, and dipole Moment are computed. A close agreement between experimental and computed bond lengths is observed from this analysis.

Chapter 4 illustrates the synergistic influence of amalgamating chitosan with dried *E. crassipes* root powder for As(III) remediation from water; later being inexpensive and abundant biomaterial. The composite was found to be very effective in removing As(III) to <10 μg/L, the permissible limit prescribed by the World Health Organization (WHO). Interactions among component materials in the composite and with adsorbed arsenic were analyzed and confirmed by different physicochemical/spectroscopic tools and further verified by DFT calculations. Physical parameters such as material dose, treatment time, and initial arsenic concentration, could alter the efficiency of the material. For a test sample containing 0.4 mg/L of arsenic, 3 g/L of the material could effectively bring down As(III) concentration below an acceptable limit. Langmuir adsorption isotherm could reasonably explain the sorption pattern and maximum adsorption capacity was found to be 7.11 mg of arsenic/g. The sorption was chemical in nature and was governed by a pseudo-second-order kinetic model.

Over the past few decades, nanomaterials have immensely enticed the scientists owing to their extraordinary properties. Association of atoms at the nanoscale level constitutes a bridge between single molecules and infinite bulk systems and imparts them unique characteristics. In the recent years, nanoporous solids and clusters have attracted great interest owing to their exceptional properties and wide applicability as both catalysts and catalytic supports in industrial processes. Advances in the scientific approach for catalyst design and developing novel functional materials through two molecular aggregations have generated interest among researchers to scrutinize the processes at the atomic level. Future development in nanotechnology significantly depends on the fundamental understandings of the structure and dynamics of nanomaterials that can be provided by multi-scale modeling and simulation. Theoretical techniques and mathematical modeling efforts are now well advanced to make reliable predictions about the properties of materials to promote their application in electronics, photonics, catalysis, sensor, and energy storage via size scaling and structural modification. Chapter 5 presents significant contributions of computational studies in the design and development of heterogeneous catalysts based on graphene and transition metal oxides.

In Chapter 6, it is intended to show that the series resistance (R_s) plays a significant role in the organic device performance. The typical value of R_s in any organic device is very high which is mainly because of the interface and the trapping of charge carriers at the bulk and also at the barrier potential at the interface. In this work, we have prepared Thionin dye-based organic diode and studied R_s by varying the dye concentration of the diode. To prepare this diode a thin layer of Thionin dye of different concentrations of 2 mg, 4 mg, 6 mg, and 8 mg are sandwiched separately in between two electrodes one of which is ITO coated glass and another is Al. The dark current-voltage (I-V) characteristics of these diodes have been studied and estimated the values of R_s by using Cheung Cheung function. It is observed that the estimated value of R_s for 2 mg is very high and by increasing the dye concentration, the value of R_s is reduced. It is also observed that the value of the ideality factor (η) also reduced by increasing the dye concentration.

In Chapter 7, we show that the Faddeev-Sominsky's process allows construct a natural inverse for any square matrix, which is an alternative to the inverse obtained by Gower. Noether's theorem is discussed in Chapter 8. The objective of Chapter 9 is to derive the problem of elastoplastic modeling of an orthotropic boron-aluminum fiber-reinforced composite thick-walled rotating cylinder subjected to a temperature gradient by using Seth's transition and generalized strain measure theory. The combined effects of temperature and angular speed have been presented numerically and graphically. Seth's transition theory does not require the assumptions: the yield criterion, the incompressibility conditions, the deformation is small, etc., and thus solves a more general problem. This theory utilizes the concept of generalized strain measure and asymptotic solution at the turning points of the differential equations defining the deformed field. It is seen that cylinders having smaller radii ratios require higher angular speed for yielding as compared to cylinders having higher radii ratios. With the inclusion of thermal effects, the angular speed increased for initial yielding to smaller radii ratio but for the fully plastic state, the angular speed is the same. It is observed that the maximum circumferential stress occurs at the internal surface for both transitional and fully plastic state at any temperature and angular speed.

In Chapter 10, elastic-plastic stress distributions in the human femur and tibia bone are calculated analytically. The bone is modeled in the form of a cylinder that exhibits orthotropic macroscopic symmetry. Seth's transition theory has been used to model the elastic-plastic state of stresses. The cylinder so modeled is subjected to external pressure. The results obtained illustrate the stress build up in the bones when there is an external force applied on them, thereby providing insights into the elastic-plastic extensions and their tendency to fracture.

In Chapter 11, we study how to generate orthogonal 4×4-matrices using a quaternionic triple product, which leads in a natural manner to Dirac matrices and the analysis of rotations in three and four dimensions.

In spite of enormous industrial applications of carbon nanostructures, the greater understanding and control of their synthesis needs to be improved based on the knowledge of carbon-carbon interactions and the interaction of carbon atoms with the other materials. The current production rate remains low, hovering in 10–20%, so an effective process of planning and design for nanomanufacturing is considered necessary for quality assurance of carbon nanostructures and consequentially the yield rate. Among the current technologies, thin films of carbon nanotubes (CNTs) represent one of the most interesting materials from an application-oriented point of view. Therefore, fundamental studies based on theoretical and numerical approaches become crucial for the design and development of high-performance devices.

Chapter 12 is focused on the basics of different carbon nanotubes (CNTs) technologies from both theoretical and an experimental point of view, for helping in the design of CNT film devices for various applications. This chapter is focused on the theoretical analysis of the networks, based on different models and levels of approximations. All the implemented theoretical models are discussed in order to explain the multi-scale approach.

CHAPTER 1

Quantum Information Perspective on Chemical Reactivity[†]

ROMAN F. NALEWAJSKI[*]

Department of Theoretical Chemistry, Jagiellonian University,
Gronostajowa 2, 30–387 Cracow, Poland,
E-mail: nalewajs@chemia.uj.edu.pl

ABSTRACT

A need for quantum information measures combining the probability and current contributions is emphasized and continuity relations are summarized. Resultant gradient information accounts for the classical (probability) and nonclassical (phase) contributions and reflects the kinetic energy content of molecular electronic state. The phase-information description enables one to distinguish the hypothetical states of the mutually bonded (entangled) and nonbonded (disentangled) reactants in donor-acceptor systems. This generalized information-theoretic perspective is applied to chemical reactivity phenomena within the *grand*-ensemble description of thermodynamic conditions in open molecular systems. The physical equivalence of variational principles for the system electronic energy and resultant gradient information is demonstrated. The virial theorem decomposition of energy profiles is used to index the Hammond rule of reactivity theory, information changes in chemical reactions are addressed, and charge transfer between reactants is explored. The

---

[†]The following notation is adopted throughout: A denotes a *scalar*, A is the row or column *vector*, A represents a square or rectangular *matrix*, and symbol A stands for the quantum-mechanical *operator* of the physical property A. The logarithm of the Shannon information measure is taken to an arbitrary but fixed base: log = log2 corresponds to the information content measured in *bits* (binary digits), while log = ln expresses the amount of information in *nats* (natural units): 1 nat = 1.44 bits.
[*]Professor Emeritus.

frontier-electron approach to molecular interactions is used to rationalize the Hard (Soft) Acid and Bases (HSAB) principle of structural chemistry, and the rule implications for the *intra*- and *inter*-reactant communications are commented upon.

1.1 INTRODUCTION

The classical Information Theory (IT) of Fisher and Shannon [1–8] has been successfully applied to interpret in chemical terms the electron probability distributions of molecular systems, e.g., [9–12]. Information principles have been explored [13–18], density pieces attributed to Atoms-in-Molecules (AIM) have been tackled [9, 13, 17–22], patterns of chemical bonds have been extracted from molecular electron communications [9–12, 23–33], entropy/information distributions in molecules have been examined [9–12, 34–37] and the nonadditive Fisher information [9–12, 36, 37] has been linked to the electron localization function (ELF) [38–40] of modern density functional theory (DFT) [41–46]. This analysis has formulated the contragradience (CG) probe [9–12, 47] for localizing chemical bonds, and the orbital communication theory (OCT) of the chemical bond [11, 12, 23–33] has identified the *bridge*-bonds originating from the cascade propagations of information between AIM, which involve intermediate orbitals [11, 12, 47–53].

In entropic theories of molecular electronic structure one ultimately requires the *quantum* generalizations of the familiar Fisher and Shannon measures of the information/entropy content in probability distributions, which are appropriate for *complex* probability amplitudes (wave functions) of quantum mechanics (QM). Probability distributions generate the *classical* entropy/information descriptors of electronic states. They reflect only the wave function *modulus*, while the wave function *phase*, or its gradient determining the *current* density, give rise to the corresponding nonclassical supplements in *resultant* IT measures, of the overall information content in molecular electronic states. In quantum information theory (QIT) [12, 54–66] the classical information terms, conceptually rooted in DFT, probe the entropic content of *incoherent* (disentangled) local "events," while their nonclassical supplements provide the nonclassical information contribution due to the mutual *coherence* (entanglement) of such local events. The resultant measures combining the probability and phase/current contributions allow one to distinguish the information content of states generating the same electron density but differing in their phase/current distributions, e.g., those characterizing the bonded and nonbonded states of polarized reactants.

The variational principles of such generalized *entropy* concepts have determined the *phase*-equilibria in molecules and their constituent fragments [12, 57–61]. The resultant *gradient*-information in a thermodynamic state of a molecule is proportional to its *ensemble*-average value of the kinetic energy of electrons [9, 12, 14, 36, 67]. This allows one to interpret the familiar variational principle for determining the minimum thermodynamic energy as equivalent *resultant*-information rule [14, 67], which forms a basis for the novel information perspective on reactivity phenomena [67].

Various DFT-based approaches to classical issues in reactivity theory use modern charge-density analysis and *energy*-centered arguments in justifying the observed reaction paths and relative yields of their products [9, 44, 68–74]. Qualitative considerations on preferences in chemical reactions usually emphasize changes in energies of reactants and of the whole reactive system, which are induced by displacements (perturbations) in parameters describing the relevant (real or hypothetical) electronic states. In such treatments, usually covering also the linear responses to these primary shifts, one explores the reactivity implications of the energy equilibrium and stability criteria [9, 11, 67–69, 72, 73]. For example, in Charge Sensitivity Analysis (CSA) [68, 69] the energy derivatives with respect to the system external potential due to the fixed nuclei (v) and its overall number of electrons (N), as well as the associated charge responses of both the whole reactive system and its constituent subsystems have been explored as potential reactivity descriptors. In the acceptor-donor complexes R ≡ A—B, combining the *acid* (A) and *base* (B) reactants, such responses can be subsequently combined into the *in situ* descriptors characterizing the B→A Charge-Transfer (CT). Such difference characteristics of the polarized subsystems can be expressed in terms of elementary (canonical) charge sensitivities of reactants [68, 69, 72].

In this perspective, we shall further explore the applicability of QIT in describing the reactivity phenomena in A—B systems. The probability and current components of electronic states will be summarized and their continuities briefly explored. The *grand*-ensemble description of thermodynamic equilibria in *externally*-open molecular systems will be used to demonstrate the physical equivalence of the energy and *resultant*-information principles. The populational derivatives of *ensemble*-averages of the electronic energy and resultant *gradient*-information in molecular equilibrium states will be shown to be fully equivalent in indexing the CT phenomena, by correctly determining both the direction and magnitude of such electron flows in reactive systems. The activation (promotion) of reactants at hypothetical stages of chemical processes, which are invoked in reactivity theory, will be examined and the *in situ* populational derivatives of the average resultant-information will be applied to determine the optimum amount of CT in donor-acceptor systems. The molecular virial theorem [75] will be used to test usefulness of the resultant-information principle in indexing and rationalizing general rules of chemical reactivity: a qualitative Hammond's postulate [76] and Pearson's Hard (Soft) Acids and Bases (HSAB) principle [77].

1.2   PROBABILITY AND CURRENT DEGREES-OF-FREEDOM OF ELECTRONIC STATES

For simplicity, let us first consider a single electron moving in the external potential $v(r)$ created by the fixed nuclei of a molecule. Its quantum state $|\psi(t)\rangle$ at time t is described by the complex wave function:

$$\psi(r, t) = \langle r|\psi(t)\rangle = R(r, t) \exp[i\phi(r, t)], \tag{1}$$

where the real functions $R(r, t) \geq 0$ and $\phi(r, t) \geq 0$ stand for its modulus and phase components, respectively. They generate the particle probability distribution:

$$p(r, t) = \psi(r, t)\,\psi(r, t)^* = R(r, t)^2, \tag{2}$$

and the current density:

$$j(r, t) = (\hbar/2mi)[\psi(r, t)^* \nabla \psi(r, t) - \psi(r, t) \nabla \psi(r, t)^*]$$

$$= (\hbar/m)\,p(r, t)\,\nabla\phi(r, t) \equiv p(r, t)\,V(r, t). \tag{3}$$

The velocity field of probability "fluid," $V(r, t) = V[r(t), t] \equiv dr(t)/dt$, measuring the local density of the current-per-particle, $V(r, t) = j(r, t)/p(r, t) = (\hbar/m) \nabla\phi(r, t)$, thus reflects the state *phase*-gradient $\nabla\phi(r, t)$. The phase component can be also expressed in terms of the wave function ratio:

$$\phi(r, t) = (2i)^{-1} \ln[\psi(r, t)/\psi(r, t)^*]. \tag{4}$$

To summarize, the *product* of complex-conjugate states in Eq. (2) and their *ratio* in Eq. (4) delineate two independent degrees-of-freedom of the particle general electronic state. They ultimately determine the complementary *physical* descriptors of the particle *probability* and *current* distributions at a specified time t:

$$\psi \Leftrightarrow \{(\psi\psi^*), (\psi/\psi^*)\} \Leftrightarrow \{R, \phi\} \Leftrightarrow \{p, j\}. \tag{5}$$

The local in homogeneities of these two fundamental fields are reflected by their gradients, which probe the complementary facets of the state *structure*-content: $\nabla p = 2R \nabla R$ extracts the spatial in homogeneity of the probability density, the static structure "of being," while $\nabla \cdot j = (\hbar/m)\nabla p \cdot \nabla \phi$ uncovers the dynamic structure of "becoming." We have used above the direct implication of the *probability*-continuity,

$$\partial p(r, t)/\partial t = -\nabla \cdot j(r, t) = - [\nabla p(r, t) \cdot V(r, t) + p(r, t) \nabla \cdot V(r, t)] \text{ or}$$
$$dp[r(t), t]/dt \equiv \sigma_p(r, t) = \partial p(r, t)/\partial t + \nabla \cdot j(r, t)$$
$$= \partial p(r, t)/\partial t + [\partial p(r, t)/\partial r] \cdot [dr(t)/dt]$$
$$= \partial p(r, t)/\partial t + \nabla p(r, t) \cdot V(r, t) = 0, \tag{6}$$

that the divergence of velocity field $V(r, t)$, determined by the state *phase*-Laplacian, identically vanishes:

$$\nabla \cdot V(r, t) = (\hbar/m) \Delta\phi(r, t) = 0. \tag{7}$$

Here, $\sigma_p \equiv dp/dt$ and $\partial p/\partial t$ denote the *total* and *partial* time-derivatives of electronic probability distribution $p(r, t) = p[r(t), t]$, thus corresponding to time rates of densities in the "moving" and "frozen" infinitesimal volume elements of the physical space, respectively. The local probability "source" corresponds to the total derivative dp/dt. It should be interpreted as time rate of change in the infinitesimal volume element of the probability fluid *flowing* with the probability current. The partial derivative $\partial p/\partial t$ represents the corresponding rate at the specified (*fixed*) point in space.

In molecular scenario one envisages the electron moving in the external potential $v(r)$, due to the "frozen" nuclei of the Born-Oppenheimer (BO) approximation, described by the electronic Hamiltonian:

$$H(r) = - (\hbar^2/2m)\nabla^2 + v(r) \equiv T(r) + v(r), \tag{8}$$

where $T(r)$ stands for its kinetic part. The dynamics of electronic state $\psi(r, t)$ is then determined by the Schrödinger equation (SE):

$$i\hbar\, \partial\psi(r, t)/\partial t = H(r)\, \psi(r, t). \tag{9}$$

This fundamental equation implies specific temporal evolutions of instantaneous distributions $\{R(r, t) \text{ or } p(r, t)\}$ and $\{\phi(r, t) \text{ or } j(r, t)\}$.

This fundamental equation implies specific temporal evolutions of instantaneous distributions $R(r, t)$ and $\phi(r, t)$. One indeed observes that substituting Eq. (1) into Eq. (9) expresses SE in terms of wavefunction components:

$$i(\partial R/\partial t) - R\,(\partial\phi/\partial t) = -[\hbar/(2m)][2i\nabla R\cdot\nabla\phi + \Delta R - R(\nabla\phi)^2] + (v/\hbar)\,R. \tag{10}$$

By equating the real and imaginary parts on both its sides ultimately gives the SE-implied dynamical equations for the modulus and phase components of molecular electronic states, respectively. A comparison of imaginary contributions gives the modulus-dynamics equation [compare probability-continuity of Eq. (6)],

$$(\partial R/\partial t) = -[(\hbar/m)\nabla\phi]\cdot\nabla R = -V\cdot\nabla R = (2R)^{-1}\,(\partial p/\partial t), \tag{11}$$

while comparing real parts predicts the following phase-dynamics:

$$\partial\phi/\partial t = [\hbar/(2m)]\,[R^{-1}\Delta R - (\nabla\phi)^2] - v/\hbar. \tag{12}$$

The total time derivative of Eq. (6) expresses a sourceless continuity relation for the probability density: $\sigma_p(r, t) = 0$. The same effective velocity, of the probability current per electron, should be also attributed to the current concept associated with the phase component: $J(r, t) = \phi(r, t)\, V(r, t)$. The phase field $\phi(r, t)$ and its current $J(r, t)$ then determine a nonvanishing *phase*-source term $\sigma_\phi(r, t)$ in the associated balance equation:

$$\sigma_\phi(r, t) \equiv d(r, t)/dt = \partial\phi(r, t)/\partial t + \nabla\cdot J(r, t) \neq 0.$$

This *phase*-continuity also expresses the *phase*-source $\sigma_\phi(r, t)$ in terms of *probability*-velocity of Eq. (3):

$$d\phi(r, t)/dt = d\phi[r(t), t]/dt = \partial\phi[r(t), t]/\partial t + dr(t)/dt \cdot \partial\phi[r(t), t]/\partial r$$

$$= \partial\phi(r, t)/\partial t + V(r, t)\cdot\nabla\phi(r, t) = \partial\phi(r, t)/\partial t + (\hbar/m)\,[\nabla\phi(r, t)]^2.$$

Using the phase-dynamics of Eq. (12) finally gives:

$$\sigma_\phi = [\hbar/(2m)]\,[R^{-1}\Delta R + (\nabla\phi)^2] - v/\hbar. \tag{13}$$

1.3 RESULTANT GRADIENT INFORMATION

At the given instant $t = t_0$, for simplicity suppressed from the list of state parameters,

$$\psi(r, t_0) = \langle r|\psi(t_0)\rangle \equiv \langle r|\psi\rangle = R(r)\exp[i\phi(r)],\, p(r, t_0) \equiv p(r),\, j(r, t_0) \equiv j(r),\, \text{etc.}$$

the average Fisher's [1, 2] measure of the *classical* gradient information for locality events contained in probability density $p(r) = R(r)^2$ is reminiscent of von Weizsäcker's [78] in homogeneity correction to the kinetic energy functional:

$$I[p] = \int [\nabla p(r)]^2/p(r)\, dr = \int p(r)\, [\nabla \ln p(r)]^2\, dr = 4\int [\nabla R(r)]^2\, dr \equiv I[R]. \quad (14)$$

The amplitude form $I[R]$ reveals that this classical descriptor measures the resultant length of the state *modulus*-gradient. This descriptor characterizes an effective "narrowness" of the particle probability distribution, i.e., a degree of determinicity in the particle position.

These probability/modulus functionals generalize naturally into the corresponding *resultant* information descriptor [12, 54–61], functional of the quantum state $|\psi\rangle$ itself, which combines the classical (probability) and non-classical (phase/current) contributions. Such generalized information concept is applicable to complex wave functions of molecular QM. It is defined by the expectation value of the Hermitian operator for the overall *gradient*-information, related to the kinetic energy operator $T(r)$ [36]:

$$I(r) = -4\Delta = (2i\nabla)^2 = (8m/\hbar^2)\,T(r) \equiv \sigma T(r). \quad (15)$$

From a straightforward integration by parts one then obtains the following expression for the resultant *gradient*-information in state $|\psi\rangle$:

$$I[\psi] = \langle \psi|I|\psi\rangle = \int |\nabla \psi(r)|^2\, dr$$

$$= I[p] + 4\int p(r)\,[\nabla \phi(r)]^2\, dr \equiv I[p] + I[\phi] \equiv I[p,\,\phi]$$

$$= I[p] + (2m/\hbar)^2 \int p(r)^{-1}\, j(r)^2\, dr \equiv I[p] + I[j] \equiv I[p,\,j]. \quad (16)$$

It also reflects the state average kinetic energy:

$$T[\psi] = \langle \psi|T|\psi\rangle = \sigma^{-1}\, I[\psi]. \quad (17)$$

This *one*-electron development can be generalized into N-electron systems in quantum state $|\Psi(N)\rangle$ exhibiting the electron density $\rho(r) =$

$Np(r)$. The corresponding information operator then combines terms due to each electron,

$$I(N) = \sum_i I(r_i) = \sigma \sum_i T(r_i) \equiv \sigma T(N), \qquad (18)$$

and determines the average gradient information,

$$I(N) = \langle \Psi(N)| I(N)|\Psi(N)\rangle = \sigma \langle \Psi(N)|T(N)|\Psi(N)\rangle = \sigma T(N), \qquad (19)$$

proportional to the expectation value $T(N)$ of the kinetic energy operator $T(N)$.

For example, in *one*-determinantal (orbital) representation of electronic states,

$$(N) = |\psi_1 \psi_2 \dots \psi_N|,$$

e.g., in the familiar Hartree-Fock of Kohn-Sham theories, these N-electron descriptors combine the additive contributions due to the (singly) occupied molecular spin-orbital's (MO)

$$\psi = (\psi_1, \psi_2, \dots, \psi_N),$$

$$T(N) = \sum_k \langle \psi_k|T|\psi_k\rangle = \sum_k T_k = \sigma^{-1} \sum_k \langle \psi_k|I|\psi_k\rangle = \sigma^{-1} \sum_k I_k \qquad (20)$$

The relevant separation of the *modulus*- and *phase*-components of such N-electron states is effected by using the Harriman-Zumbach-Maschke (HZM) [79, 80] construction of electronic states yielding the prescribed electron distribution [43]. It uses the (complex) Equidensity Orbital's (EO), all yielding the molecular probability distribution $p(r)$, and exhibiting the density-dependent local phases safeguarding the MO mutual orthogonality.

Similar *resultant* concepts of the electronic global entropy have been also introduced [12, 54]. In the simplest, *one*-electron case they combine the classical Shannon's [3, 4] entropy in electronic probability distribution:

$$S[p] = - \int p(r) \ln p(r) \, dr,$$

and nonclassical contribution [12, 54–61] reflecting the average phase,

$$S[\phi] = - 2 \int p(r) \phi(r) \, dr,$$

$$S[p, \phi] = S[p] + S[\phi] = \langle \psi|-\ln p - 2\phi|\psi\rangle \equiv \langle \psi|S(r)|\psi\rangle. \qquad (21)$$

In the *complex*-entropy concept [12, 54], the expectation value of the (*non*-Hermitian) entropy operator

$$S(r) = -2\ln\psi(r) = -2[\ln R(r) + i\phi(r)] = -\ln p(r) - 2i\phi(r)],$$

these two contributions constitute the real and imaginary components of such *state*-descriptor:

$$S[\psi] = \langle\psi|-2\ln\psi|\psi\rangle \equiv \langle\psi|S|\psi = S[p] + iS[\phi], \tag{22}$$

Since the nonclassical gradient information $I[\phi] = I[j]$ in the resultant *determinicity* measure $I[p, \phi] = I[p,j]$ is positive, its *indeterminicity* analog $M[\phi] \equiv -I[\phi] = M[j] \equiv -I[j]$ in the resultant gradient entropy [12]

$$M[p, \phi] \equiv I[p] - I[\phi] \equiv M[p] + M[\phi] \equiv I[p] - I[j] \equiv M[p] + M[j] \tag{23}$$

must be negative. Indeed, a nonvanishing current pattern contributes an extra information ("order") contribution into the resultant IT descriptor of molecular electronic structure, thus decreasing the overall level of the system gradient entropy ("disorder") contribution.

The *state*-extremum of the resultant ("*scalar*") global-entropy measure of Eq. (21),

$$\delta S[\delta\psi^*] = \langle\delta\psi|-\ln p - 2\phi|\psi\rangle|_{eq.} = 0,$$

predicts the equilibrium phase

$$\phi_{eq.}(r) = -(1/2)\ln p(r) \geq 0. \tag{24}$$

This entropy-optimum phase generates the associated equilibrium current in the direction opposite to the probability gradient:

$$j_{eq.}(r) = (\hbar/m)p(r)\nabla\phi_{eq.}(r) = -\hbar/(2m)\nabla p(r).$$

The same solution marks the extremum of the resultant *gradient*-entropy of Eq. (23):

$$\delta M[\delta\psi^*] = \langle\delta\psi|(\nabla\ln p)^2 - 4(\nabla\phi)^2|\psi\rangle|_{eq.} = 0 \text{ or}$$

$$(\nabla\ln p)^2 - (2\nabla\phi_{eq.})^2 = \nabla(\ln p - 2\phi_{eq.})\cdot\nabla(\ln p + 2\phi_{eq.}) = 0.$$

A nonnegative solution (phase convention) then identifies the equilibrium phase of Eq. (24).

The proportionality relations between the average gradient-information and kinetic-energy functionals allow one to interpret the electronic energy principles as the physically equivalent *resultant*-information rules [9–12, 36, 67]. Thus the familiar variational principle of QM for the minimum of the system electronic energy

$$E(N) = \langle \Psi(N)|H(N)|\Psi(N)\rangle = T(N) + [V(N) + U(N)] \equiv T(N) + W(N),$$

where $W(N) = \langle \Psi(N)|W(N)|\Psi(N)\rangle = V(N) + U(N)$ stands for the system overall potential energy combining the electron attraction $V(N)$ and repulsion $U(N)$ contributions,

$$\min_{\Psi(N)}[E(N) - \lambda\langle\Psi(N)|\Psi(N)\rangle],$$

also represents the *potential*-energy constrained rule for the resultant *gradient*-information:

$$\min_{\Psi(N)}[I(N) - \omega\,W(N) - \kappa\langle\Psi(N)|\Psi(N)\rangle]. \tag{25}$$

Here the system Coulombic Hamiltonian of N-electron system, $H(N)$ $= h(N) + U(N)$, contains the sum $h(N)$ of one-electron operators, $h(N) = \sum_i [T(r_i) + v(r_i)] \equiv \sum_i H(r_i) = T(N) + V_{e,n}(N)$, and of the sum of electron-repulsion contributions, $U(N) = \sum_i\sum_{j>i} g(r_{i,j})$, $H(N) = h(N) + U(N)$. Combining the electronic potential energy $V_{e,n}(N) + U(N) \equiv V(N)$ and the nuclear repulsion energy $V_{n,n}$ then defines the system total potential energy $W(N) = V(N) + V_{n,n}$, which is constrained in Eq. (25), where the Lagrange multipliers κ and ω enforce the wavefunction normalization and the prescribed value of the overall potential energy.

To summarize, the modulus (probability) and phase (current) components of electronic states are both accounted for in the resultant measures of the gradient or global descriptors of the information/entropy content for generally complex wavefunctions of molecular QM. These generalized descriptors combine the familiar classical functionals of the system probability density and relevant nonclassical supplements due to the state phase or current. Their densities-per-electron satisfy classical relations linking the gradient and global information/entropy descriptors, appropriately generalized for complex electronic states [12, 54]. The Hermitian information operator $I(r)$ gives rise to real MO expectation value of the state resultant determinicity-information content $I[\psi]$, related to the average orbital kinetic energy $T[\psi]$, while the *non*-Hermitian entropy operator $S(r)$ generates complex expectation value $S[\psi]$. The proportionality relation

$I[\psi] = \sigma\, T[\psi]$ allows one to cast the familiar variational principle for the minimum-energy as the associated potential energy constrained information rule (see also Section 1.6).

1.4 INFORMATION CONTINUITY

In molecular QM the electronic wavefunctions are determined by SE. As we have argued in Section 1.2, SE also determines the dynamics equations for the state modulus and phase components, summarized by the associated continuity equations. The sourceless probability-continuity have been shown to be accompanied by the phase-balance equation exhibiting a finite local production term. Again, for simplicity reasons, in what follows we shall examine a single electron in the complex state of Eq. (1).

The state resultant-information concept $I[\psi(t)]$ [Eq. (16)], which reflects the dimensionless kinetic energy $T[\psi(t)]$, combines the classical (probability) contribution of Fisher and nonclassical (phase/ current) supplement $I[\phi; t] = I[j; t]$,

$$I[\psi(t)] \equiv I(t) = I[p, \phi; t] \equiv I[p; t] + I[\phi; t] \equiv \int p(r, t)\, I(r, t)\, dr$$

$$= I[p, j; t] \equiv I[p; t] + I[j; t] = \sigma\, T[\psi(t)].$$

Elsewhere a temporal evolution of this integral measure of the overall information content in the specified quantum state has been examined [11,12,66]. In Schrödinger's dynamical picture the time change of the resultant gradient information, the operator of which does not depend on time explicitly, results solely from the time dependence of the state vector itself. The *time*-derivative of this average information measure is thus generated by the expectation value of the commutator

$$[\mathrm{H}, \mathrm{I}] = [v, \mathrm{I}] = 4[\nabla^2, v] = 4\{[\nabla, v]\cdot\nabla + \nabla\cdot[\nabla, v]\},\ [\nabla, v] = \nabla v,$$

$$dI(t)/dt = (i/\hbar)\langle\psi(t)|[\mathrm{H}, \mathrm{I}]|\psi(t)\rangle. \tag{26}$$

This integral production of resultant gradient-information descriptor thus reads:

$$\sigma_I(t) \equiv dI(t)/dt = -(8/\hbar)\int p(r, t)\, \nabla\phi(r, t)\cdot\nabla v(r)\, dr = \sigma\int j(r, t)\cdot F(r)\, dr, \tag{27}$$

where $F(r) = -\nabla v(r)$ denotes the force due to the system fixed nuclei. This information derivative is seen to be determined by thermodynamic-like product of this external force ("affinity") and probability current ("flux"), in close analogy to the entropy source in irreversible thermodynamics [81]. The information "source" is seen to be of purely nonclassical (phase/current) origin: it is absent when the local component of the state phase and hence the state electronic current identically vanish.

This conclusion can be also demonstrated directly:

$$\sigma_I(t) \equiv dI(t)/dt = dI[p;t]/dt + dI[\phi;t]/dt = dI[\phi;t]/dt,$$

since by the (sourceless) probability continuity $dp/dt = 0$ and hence

$$dI[p;t]/dt = \int [dp(\mathbf{r},t)/dt]\, \{\delta I[p;t]/\delta p(\mathbf{r},t)\}\, d\mathbf{r} = 0 \text{ and}$$

$$\int [dp(\mathbf{r},t)/dt]\, \{\delta I[\phi;t]/\delta p(\mathbf{r},t)\}\, d\mathbf{r} = 0.$$

Therefore, the integral source of resultant *gradient*-information $I(t)$ in fact reflects the total time-dependence of its nonclassical contribution $I[\phi; t]$ through the state-phase itself,

$$dI[\phi;t]/dt = \int [d\phi(\mathbf{r},t)/dt]\, \{\delta I[\phi;t]/\delta \phi(\mathbf{r},t)\}\, d\mathbf{r}$$

$$= -8\int \sigma_\varphi(\mathbf{r},t)\, \nabla p(\mathbf{r},t) \cdot \nabla \phi(\mathbf{r},t)\, d\mathbf{r} = 8\int p(\mathbf{r},t)\, \nabla \sigma_\varphi(\mathbf{r},t) \cdot \nabla \phi(\mathbf{r},t)\, d\mathbf{r}, \qquad (28)$$

where the last expression follows from the preceding one via the integration by parts and by taking into account the vanishing divergence of velocity field [Eq. (7)].

The *global* information source

$$\sigma_I(t) = dI(t)/dt = \int [\sigma_p(\mathbf{r},t)\, I(\mathbf{r},t) + p(\mathbf{r},t)\, \sigma_I(\mathbf{r},t)]\, d\mathbf{r} = \int p(\mathbf{r},t)\, \sigma_I(\mathbf{r},t)\, d\mathbf{r},$$

can be also interpreted in terms of the *local* continuity relation for the density-per-electron $I(\mathbf{r},t)$ of the resultant gradient information,

$$\sigma_I(\mathbf{r},t) = dI(\mathbf{r},t)/dt = \partial I(\mathbf{r},t)/\partial t + \nabla \cdot \mathbf{J}_I(\mathbf{r},t), \qquad (29)$$

where $\mathbf{J}_I(\mathbf{r},t) = I(\mathbf{r},t)\, \mathbf{V}(\mathbf{r},t)$ stands for the local information current. Taking into account Eq. (7) and the continuity relations for the probability and phase distributions,

$$\partial p(\mathbf{r},t)/\partial t = -\mathbf{V}(\mathbf{r},t) \cdot \nabla p(\mathbf{r},t) = -(\hbar/m)\, \nabla \phi(\mathbf{r},t) \cdot \nabla p(\mathbf{r},t) \text{ or}$$

$$\partial R(\mathbf{r},t)/\partial t = 2R(\mathbf{r},t)^{-1}\, \partial p(\mathbf{r},t)/\partial t = -(\hbar/m)\, \nabla \phi(\mathbf{r},t) \cdot \nabla R(\mathbf{r},t) \text{ and}$$

$$\partial \phi(\mathbf{r},t)/\partial t = -\nabla \cdot \mathbf{J}(\mathbf{r},t) + \sigma_\varphi(\mathbf{r},t) = -\mathbf{V}(\mathbf{r},t) \cdot \nabla \phi(\mathbf{r},t) + \sigma_\varphi(\mathbf{r},t) = -(\hbar/m)\, [\nabla \phi(\mathbf{r},t)]^2,$$

then indeed reconstructs the integral information source of Eq. (28). The explicit terms in Eq. (29) read:

$$\partial I(\mathbf{r}, t)/\partial t = g(\mathbf{r}, t) + 8\nabla \phi(\mathbf{r}, t) \cdot \nabla \sigma_\varphi(\mathbf{r}, t) \text{ and}$$
$$\nabla \cdot \mathbf{J}_I(\mathbf{r}, t) = \nabla I(\mathbf{r}, t) \cdot \mathbf{V}(\mathbf{r}, t) = -g(\mathbf{r}, t),$$

$$g(\mathbf{r}, t) = 2\{[\nabla \ln p(\mathbf{r}, t)]^2 - p(\mathbf{r}, t)^{-1} \Delta p(\mathbf{r}, t)\} \nabla \ln p(\mathbf{r}, t) \cdot \mathbf{V}(\mathbf{r}, t).$$

The gradients of the wavefunction components determine its time evolution. Their dynamical (current) aspect is embodied in the effective velocity of the probability "fluid," which is determined by the *phase*-gradient alone. Therefore, to paraphrase Progogine [82], the gradients of the state components describe the dynamic "structure of becoming," while the instantaneous probability and phase distributions themselves characterize the state static "structure of being." The above continuity relations testify that the former determines the time evolution of the latter.

To summarize, an inclusion of the phase/current component of the resultant gradient information generates a nonvanishing source of this (dimensionless) kinetic-energy descriptor. The integral production of the nonclassical information assumes thermodynamic-like form, of the product of electronic current ("flux") and external force ("affinity"). It can be also understood in terms of the local continuity relation, for the density-per-electron of the overall gradient information. Such a local continuity framework employs the balance equations for the wavefunction modulus (probability) and phase (current) components, both implied by SE, with the information-flux and phase-current being driven with the same quantum-mechanical velocity of the probability current.

To conclude this section we briefly examine implications of the asymptotic decay of molecular electron density [83] for the equilibrium phase of Eq. (24). One recalls, that in distances r from a molecule M large compared to interatomic distances the decay of electronic density $\rho(r)$, and hence also of its probability factor $p(r) = \rho(r)/N$, are both determined by the system ionization potential $I = -\varepsilon$, negative energy ε of the highest occupied Kohn-Sham orbital:

$$\rho(r) \rightarrow \exp(-4Ir), \quad r \rightarrow \infty. \tag{30}$$

This asymptotic behavior of the density thus relates the equilibrium phase to this familiar energy descriptor,

$$\phi^{eq}(\mathbf{r}) = -(1/2)\ln p(\mathbf{r}) \propto I_M r. \tag{31}$$

The associated current,

$$j^{eq.}(r) = -\hbar/(2m)\nabla p(r) \propto I(r/r),\tag{32}$$

thus predicts the magnitude of the probability velocity to be proportional to the system ionization potential:

$$V^{eq.}(r) \propto I p(r)^{-1}(r/r).$$

Therefore, in accordance with an elementary physical intuition, in the chemically hard systems exhibiting high ionization potentials electrons move "fast," while they move "slow" in the chemically soft systems characterized by low ionization energies.

1.5 VIRIAL THEOREM DECOMPOSITION OF ENERGY PROFILES

Elsewhere [67] a direct application of the overall *gradient*-information principle to the qualitative Hammond [76] postulate of chemical reactivity theory has been reported. For a bimolecular reactive system R, A + B → R‡ → C + D, this qualitative rule emphasizes a relative resemblance of the reaction Transition-State (TS) complex R‡ to its substrates (products) in the *exoergic* (*endoergic*) reactions, while for the vanishing reaction energy the position of TS complex is predicted to be located symmetrically between substrates and products (Figure 1.1). The activation barrier thus appears "early" in *exoergic* reactions, e.g., H$_2$ + F → H + HF, with the substrates being only slightly modified in TS, R$^{\ddagger} \approx$ [A—B], both electronically and geometrically. Accordingly, in *endoergic* bond-breaking–bond-forming process, e.g., H + HF → H$_2$ + F, the barrier is "late" along the reaction-progress axis P, the *arc*-length of the Reaction Coordinate (RC) Q_c, and the activated complex resembles more products, R$^{\ddagger} \approx$ [C—D]. This qualitative statement has been subsequently given several more quantitative formulations and physical explanations using both the energetic and entropic arguments [9, 75, 84–91].

One recalls that the virial theorem in BO approximation allows one to decompose differences in the ground-state energies (E) into the conjugate kinetic (T) and potential (W) components [75]. The energy profile along the reaction "progress" coordinate P, $\Delta E(P) = E(P) - E(P_{sub.})$, is thus directly "translated" into the associated displacement in its kinetic-energy contribution, $T(P) = T(P) - T(P_{sub.})$, proportional to the corresponding

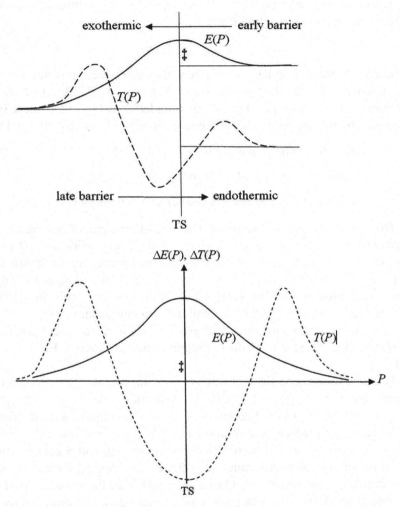

FIGURE 1.1 Variations of the electronic total (E) and kinetic (T) energies in exo-ergic ($\Delta E_r < 0$), endo-ergic ($\Delta E_r > 0$) (upper Panel) reactions, and on the symmetrical PES ($\Delta E_r = 0$) (lower Panel).

change $\Delta I(P) = I(P) - I(P_{sub.})$ in the system resultant *gradient*-information $\Delta I(P) = \sigma \Delta T(P)$:

$$\Delta T = -\Delta E - P\,[d\Delta E/dP] = -\,d[P\Delta E]/dP. \tag{33}$$

The reaction energy profile $\Delta E(P)$ in the *endo-* or *exo*-direction, for the positive and negative reaction energy

$$\Delta E_r = E(P_{prod.}) - E(P_{sub.}), \tag{34}$$

respectively, thus uniquely determines the associated profiles of the kinetic-energy or resultant-information: $\Delta T(P) \propto \Delta I(P)$. A reference to qualitative plots in Figure 1.1 shows that the latter distinguishes these two directions by the sign of the kinetic-energy/information derivative at TS:

$$endo\text{-direction: } \{(dI/dP)_{\ddagger} > 0 \text{ and } (dT/dP)_{\ddagger} > 0, \Delta E_r > 0; \tag{35}$$

$$\text{energy-}neutral\text{: } \{(dI/dP)_{\ddagger} = 0 \text{ and } (dT/dP)_{\ddagger} = 0, \Delta E_r = 0; \tag{36}$$

$$exo\text{-direction: } \{(dI/dP)_{\ddagger} < 0 \text{ and } (dT/dP)_{\ddagger} < 0, \Delta E_r < 0. \tag{37}$$

This observation demonstrates that RC-derivative of the resultant *gradient*-information at the TS complex, $dI/dP|_{\ddagger}$, proportional to $dT/dP|_{\ddagger}$, can serve as an alternative detector of the reaction energetic character: its positive (negative) values respectively identify the *endo* (*exo*) ergic reactions exhibiting the late (early) activation barriers, with the neutral case, for $\Delta E_r = 0$ and $dT/dP|_{\ddagger} = 0$, exhibiting an equidistant position of TS between the reaction substrates and products on a symmetrical potential energy surface (PES), e.g., in the hydrogen exchange reaction $H + H_2 \rightarrow H_2 + H$.

The reaction energetical character also implies its net information displacement $\Delta I_r = I(P_{prod.}) - I(P_{sub.})$, proportional to $\Delta T_r = T(P_{prod.}) - T(P_{sub.}) = -\Delta E_r$. This virial theorem association thus implies a net decrease of the resultant gradient information in *endo*-ergic processes, $\Delta I_r(endo) < 0$, its increase in *exo*-ergic reactions, $\Delta I_r(exo) > 0$, and a conservation of the overall *gradient*-information in the *energy*-neutral chemical rearrangements: $\Delta I_r(neutral) = 0$. One also recalls that the classical part of this information displacement probes an average spatial inhomogeneity of the electronic density. Therefore, the *endo*-ergic processes, requiring a net supply of energy to R, give rise to relatively less compact electron distributions in the reaction products, compared with those in the substrates. Accordingly, the *exo*-ergic transitions, which net release the energy from R, generate a relatively more concentrated electron distributions in products compared to substrates, and no such an average change is predicted in the energy-*neutral* case. This resultant information analysis accords with the virial theorem study of the origins of the covalent chemical bond by Ruedenberg at al. [92–95].

1.6 ENSEMBLE DESCRIPTION OF THERMODYNAMIC CONDITIONS

Let us briefly summarize the *grand*-ensemble basis of all quantum populational derivatives of the energy/information descriptors of the *externally*-open molecular systems, including those of the (constrained) *ensemble*-averages of electronic energy and resultant *gradient*-information [44, 68, 96, 97]. Indeed, only the average overall number of electrons $\langle N \rangle_{ens.} \equiv \mathcal{N}$ in the *externally*-open molecular part $\langle M(v) \rangle_{ens.}$, identified by the system external potential v, of the equilibrium combined (macroscopic) system $\mathcal{M} = (\langle M(v) \rangle_{ens.} | \mathcal{R})$ including the electron reservoir $\mathcal{R}$,

$$\mathcal{N} = \mathrm{tr}(DN) = \Sigma_i N_i (\Sigma_j P_j^i) \equiv \Sigma_i N_i P^i, \ \Sigma_i P^i = 1, \tag{38}$$

exhibits a continuous (fractional) spectrum of values, thus justifying the very concept of the populational ($\mathcal{N}$) derivative itself. Here, the density operator $D = \Sigma_i \Sigma_j |\psi_j^i\rangle P_j^i \langle \psi_j^i|$ identifies the statistical mixture of the system (pure) stationary states $\{|\psi_j^i\rangle \equiv |\psi_j(N_i)\rangle\}$ for different (integer) numbers of electrons $\{N_i\}$, the eigenstates of electronic Hamiltonians $\{H(N_i, v)\}$:

$$H(N_i, v)|\psi_j^i\rangle = E_j^i |\psi_j^i\rangle, j = 0, 1, \ldots,$$

which appear with (external) probabilities $\{P_j^i\}$ in the ensemble, and $N = \Sigma_i N_i (\Sigma_j |\psi_j^i\rangle\langle\psi_j^i|)$ stands for the particle number operator in Fock's space. Such $\mathcal{N}$-derivatives are involved in definitions of the system reactivity criteria, e.g., its chemical potential (negative electronegativity) [44, 68, 96–100] or the chemical hardness (softness) [44, 101] and Fukui Function (FF) [44, 102] descriptors.

The $\mathcal{N}$-derivatives are thus definable only in the system *mixed* electronic states, e.g., those corresponding to the thermodynamic equilibrium in the *externally*-open molecule $\langle M(v) \rangle_{ens.}$. In the *grand*-ensemble this state is determined by the equilibrium density operator $D_{eq.} \equiv D(\mu, T; v)$ specified by the corresponding thermodynamic (*externally*-imposed) *intensive* parameters: the chemical potential of electron reservoir, $\mu = \mu_{\mathcal{R}}$, and the absolute temperature T of heat bath, $T = T_{\mathcal{B}}$. They determine the relevant Legendre-transform of the ensemble-average energy,

$$\mathcal{E}[D] \equiv \mathcal{E}(\mathcal{N},S;v) = \text{tr}(DH) = \sum_i \sum_j P_j^i E_j^i,$$

called the *grand*-potential,

$$\Omega = \mathcal{E} - (\partial\mathcal{E}/\partial\mathcal{N})\,\mathcal{N} - (\partial\mathcal{E}/\partial S)\,S = \mathcal{E} - \mu\,\mathcal{N} - TS, \tag{39}$$

which corresponds to replacing the (*ensemble*-average) "extensive" state parameters, the particle number $\mathcal{N}$ and thermodynamic entropy:

$$S[D] = \text{tr}(DS) = \sum_i \sum_j P_j^i S_j^i(P_j^i)\}, \ S_j^i(P_j^i) = -k_B \ln P_j^i,$$

where k_B denotes the Boltzmann constant, by their respective "intensive" conjugates reflecting applied thermodynamic conditions: the ensemble chemical potential $\mu = \mu_\mathcal{R}$ and the absolute temperature $T = T_\mathcal{B}$.

The thermodynamic potential of Eq. (39) includes these external intensities as Lagrange multipliers enforcing constraints of specified values of the system average particle-number, $\langle N \rangle_{\text{ens.}} = \mathcal{N}$, and its thermodynamic-entropy, $\langle S \rangle_{\text{ens.}} = S$, at the *grand*-potential minimum:

$$\min_D \Omega[D] = \Omega[D(\mu, T; v)] \equiv \Omega(\mu, T; v). \tag{40}$$

The chemical potential level of the macroscopic reservoir is (externally) controlled to give the prescribed value of the *ensemble*-average number of electrons $\mathcal{N}$, $\mu = \mu(\mathcal{N})$. The externally imposed parameters (μ, T) ultimately determine the optimum probabilities of the (pure) stationary states (eigenstates of the electronic Hamiltonians for integer electron numbers), which describe the mixed (equilibrium) state in the *grand*-ensemble:

$$P_j^i(\mu, T; v) = \Xi^{-1} \exp[\beta(\mu N_i - E_j^i)], \tag{41}$$

and the associated density operator:

$$D(\mu, T; v) = \sum_i \sum_j |\psi_j^i\rangle P_j^i(\mu, T; v) \langle\psi_j^i| \equiv D_{\text{eq.}}; \tag{42}$$

here, Ξ stands for the *grand*-ensemble partition function and $\beta = (k_B T)^{-1}$. The equilibrium probabilities of Eq. (41) represent eigenvalues of the *grand*-canonical statistical operator acting in Fock's space:

$$d(\mu, T; v) = \exp\{\beta[\mu N - H(v)]\}/\text{tr}(\exp\{\beta[\mu N - H(v)]\}). \tag{43}$$

Consider the *grand*-ensemble representing the *externally*-open molecule $\langle M(v) \rangle_{\text{ens.}}$ at the *zero*-temperature limit, $T \to 0$, coupled to the electron reservoir $\mathcal{R}$ in the combined system $\mathcal{M}$. This mixture of molecular *ground*-states $\{|\psi_0^i\rangle = \psi_0[N_i, v]\}$, defined for the integer number of electrons $N_i =$

$\langle \psi_0^i|N|\psi_0^i\rangle$ and corresponding to energies $E_0^i = \langle \psi_0^i|H(N_i, v)|\psi_0^i\rangle = E_0[N_i, v]$, which appear in the ensemble with the equilibrium thermodynamic probabilities $\{P_0^i(\mu, T{\to}0; v)\}$, represents the lowest equilibrium state of an *externally*-open molecule $\langle M(\mu, T{\to}0; v)\rangle_{ens.}$ in such *externally*-imposed thermodynamic conditions. In this $T{\to}0$ limit only two *ground*-states ($j = 0$), $|\psi_0^i\rangle$ and $|\psi_0^{i+1}\rangle$, corresponding to neighboring integers "bracketing" the given (fractional) $\langle N\rangle_{ens.} = \mathcal{N}$, $N_i \leq \mathcal{N} \leq N_i + 1$, appear in the equilibrium *mixed*-state [44, 97]. Their ensemble probabilities for the specified average electron number

$$\langle N\rangle_{ens.} = i\, P_i(T{\to}0) + (i + 1)[1 - P_i(T{\to}0)] = \mathcal{N}$$

read:

$$P_i(T{\to}0) = 1 + i - \mathcal{N} \equiv 1 - \omega \text{ and } P_{i+1}(T{\to}0) = \mathcal{N} - i \equiv \omega.$$

The continuous $\mathcal{N}$–section of the energy function $\mathcal{E}(\mathcal{N},S)$ then consists of the *straight*-line segments between the neighboring integer values of $\mathcal{N}$. This implies constant values of the chemical potential in all such admissible (partial) ranges of the system average number of electrons, and μ-discontinuity at the integer values of the average electron-number $\{\mathcal{N} = N_i\}$.

In this thermodynamic scenario the ensemble probabilities of *pure*-states in the mixture thus result from the variational principle [Eq. (40)] for the thermodynamic potential of Eq. (39):

$$\Omega[D] = \mathcal{E}[D] - \mu\,\mathcal{N}[D] - T\,S[D] = \mathrm{tr}(D\Omega),$$

$$\Omega(\mu, T; v) = H(v) - \mu N - TS, \quad S = -k_B \ln d, \tag{44}$$

$$\min_D \Omega[D] = \Omega[D(\mu, T; v)] = \mathcal{E}[D_{eq.}] - \mu\,\mathcal{N}[D_{eq.}] - T\,S[D_{eq.}]. \tag{45}$$

The relevant ensemble averages of the system energy and its electron number read

$$\langle E(\mu, T)\rangle_{ens.} = \mathcal{E}[D_{eq.}] = \sum_i \sum_j P_j^i(\mu, T; v)\, E_j^i,$$

$$\langle N(\mu, T)\rangle_{ens.} = \mathcal{N}[D_{eq.}] = \sum_i [\sum_j P_j^i(\mu, T; v)]\, N_i = \sum_i P^i(\mu, T; v)\, N_i, \tag{46}$$

and von Neumann's [103] *ensemble*-entropy:

$$\langle S(\mu, T)\rangle_{ens.} = S[D_{eq.}] = \mathrm{tr}(D_{eq.}S) = -k_B\, \mathrm{tr}(D_{eq.} \ln D_{eq.})$$

$$= -k_B \sum_i \sum_j P_j^i(\mu, T; v) \ln P_j^i(\mu, T; v) \equiv \sum_i \sum_j P_j^i(\mu, T; v) S_j^i[P_j^i(\mu, T; v)],$$

$$S(\mu, T; v) = -k_B \sum_i \sum_j |\psi_j^i\rangle \ln P_j^i(\mu, T; v) \langle \psi_j^i|,$$

$$S_j^i[P_j^i(\mu, T; v)] = \langle \psi_j^i|S(\mu, T; v)|\psi_j^i\rangle = -k_B \ln P_j^i(\mu, T; v). \tag{47}$$

One observes that the latter identically vanishes in the *pure* quantum state $|\psi_j^i\rangle$, when $P_j^i = 1$ for the vanishing probabilities of the remaining stationary states.

The minimum value of the grand potential also identifies thermodynamic values of the two Lagrange multipliers in this principle as partial derivatives of the ensemble average energy $\mathcal{E}[D_{eq.}] = \mathcal{E}[\mu, T; v]$ with respect to the corresponding constraint values. In the bound electronic state:

$$\mu = (\partial \mathcal{E}/\partial \mathcal{N}) < 0 \text{ and } T = (\partial \mathcal{E}/\partial S) > 0. \tag{48}$$

The associated average value of the resultant gradient information, given by the weighted expression in terms of the equilibrium probabilities in this thermodynamic *mixed*-state:

$$\langle I \rangle_{ens.} = \mathcal{J}[D_{eq.}] = tr(D_{eq.}I) = \sum_i \sum_j P_j^i(\mu, T; v) \langle \psi_j^i|I|\psi_j^i\rangle \equiv \sum_i \sum_j P_j^i(\mu, T; v) I_j^i,$$

$$I_j^i = \sigma \langle \psi_j^i|T|\psi_j^i\rangle \equiv \sigma T_j^i, \tag{49}$$

is related to the ensemble average kinetic energy:

$$\langle T \rangle_{ens.} \equiv \mathcal{T} = tr(DT) = \sum_i \sum_j P_j^i(\mu, T; v) \langle \psi_j^i|T|\psi_j^i\rangle = \sum_i \sum_j P_j^i(\mu, T; v) T_j^i = \sigma^{-1}\mathcal{J}. \tag{50}$$

Therefore, the thermodynamic rule of Eqs. (40) and (45), for the constrained minimum of the ensemble-average electronic energy, can be alternatively interpreted as the corresponding principle for the constrained overall *gradient*-information content [36, 67]:

$$\sigma \min_D \Omega[D] = \sigma \Omega[D_{eq.}] = \mathcal{J}[D_{eq.}] + \sigma\{W[D_{eq.}] - \mu \mathcal{N}[D_{eq.}] - TS[D_{eq.}]\}. \tag{51}$$

Here, the *ensemble*-average value of the system overall potential energy,

$$W[D_{eq.}] = \mathcal{V}[D_{eq.}] + \mathcal{U}[D_{eq.}], \tag{52}$$

combines the average nuclear attraction ($\mathcal{V}[D_{eq.}]$) and repulsion ($\mathcal{U}[D_{eq.}]$) contributions.

The information principle of Eq. (51) is seen to contain an additional constraint of the fixed potential energy, $\langle W \rangle_{ens.} = \mathcal{W}$, multiplied by the Lagrange multiplier:

$$\lambda_W = -\sigma = (\partial \mathcal{I}/\partial \mathcal{W})_{N,S}\big|_{eq.} \equiv \omega < 0. \tag{53}$$

It also includes the modified "intensities" associated with the remaining constraints:

$$\text{information } potential: \lambda_N = \sigma\mu = (\partial \mathcal{I}/\partial N)_{\mathcal{W},S}\big|_{eq.} \equiv \xi < 0 \text{ and} \tag{54}$$

$$\text{and information } "temperature": \lambda_S = \sigma T = (\partial \mathcal{I}/\partial S)_{\mathcal{W},N}\big|_{eq.} \equiv \tau > 0. \tag{55}$$

It should be stressed that the conjugate thermodynamic principles, for the constrained extrema of the *ensemble*-average energy:

$$\delta(\mathcal{E}[D] - \mu N[D] - TS[D])\big|_{eq.} = 0, \tag{56}$$

and of *thermodynamic*-mean *gradient*-information:

$$\delta(\mathcal{I}[D] - \omega \mathcal{W}[D] - \xi N[D] - \tau S[D])\big|_{eq.} = 0, \tag{57}$$

have the same optimum probability solutions of Eq. (41). This manifests the physical equivalence of the energetic and entropic principles in determining equilibria in ordinary thermodynamics [81].

The same ensemble interpretation applies to the diagonal and mixed *second* derivatives of the average electronic energy, which involve differentiation with respect to electron-population variable N. In energy representation the *chemical hardness* reflects N-derivative of the chemical potential:

$$\eta = \partial^2 \mathcal{E}/\partial N^2 = \partial\mu/\partial N > 0,$$

while the *information hardness* [67] measures the N-derivative of information potential:

$$\theta = \partial^2 \mathcal{I}/\partial N^2 = \partial\xi/\partial N = \sigma\eta > 0. \tag{58}$$

By Maxwell's *cross*-differentiation relation the mixed second derivative of the system energy

$$f(r) = \partial^2 \mathcal{E}/\partial N \, \partial v(r) = \delta\mu/\delta v(r) = \partial\rho(r)/\partial N,$$

measuring the global Fukui-Function (FF), can be alternatively interpreted as either the density response per unit populational displacement or the global chemical potential response to unit local change in the system external potential. The associated mixed derivative of the average resultant *gradient*-information then reads [see Eq. (54)]:

$$\varphi(r) = \partial^2 \mathcal{J}/\partial \mathcal{N} \, \partial v(r) = \delta \xi / \delta v(r) = \sigma f(r), \tag{59}$$

The positive signs of the diagonal second derivative descriptors assure the external *stability* of $\langle M(v) \rangle_{ens.}$ with respect to hypothetical electron flows between the molecular system and its reservoir. Indeed, they imply an increase (a decrease) of the global energetic and information "intensities" μ and ξ, coupled to $\mathcal{N}$, in response to the perturbation created by the primary electron inflow (outflow). This is in accordance with the familiar Le Châtelier and Le Châtelier-Braun principles of thermodynamics [81], that those spontaneous responses in the system intensities to the initial population displacements diminish effects of the primary perturbations.

In *energy*-representation the SE or equivalently the *minimum*-energy principle of QM *determines* the system ground-state $|\psi_0[N, v]\rangle = |\psi_0[\rho_0]\rangle$, the unique functional of its electron density $\rho_0(r) = \rho[\mathcal{N}, v; r] = |\psi_0(r)|^2$, the equilibrium distribution for N-electron Hamiltonian $H(N, v)$. The overall electronic energy, is then determined by the density functional [41–46]:

$$E_v[\rho_0] = \int \rho_0(r) \, v(r) \, dr + F[\rho_0] + U_{n,n}, \, F[\rho_0] = \langle \psi_0[\rho_0] | F(N) | \psi_0[\rho_0] \rangle, \tag{60}$$

where $F(N) = T(N) + U(N)$. The *entropy*-representation of such individual (closed) *microscopic* systems subsequently offers an IT *interpretation* of this fundamental energetical result. This is contrary to the ordinary thermodynamics of *macroscopic* systems [81], where the familiar principles of the *minimum*-energy (for constant entropy) and of the *maximum*-entropy (for constant energy) are *both* capable of the precise identification of the system equilibrium state.

Therefore, application of QIT to individual molecular systems in their *pure* quantum states generates only an alternative interpretation tool of the closed-system electron distribution, determined from the fundamental energetic (SE) calculations. It employs the IT *entropic* concepts and methods to explain the information content of the known (quantum) electronic structure of molecules, extract their bonding patterns and reactivity preferences, and search for thermodynamic-like relations and analogies in molecular scenarios. This distinguishes this supplementary outlook on the quantum distribution of electrons in molecules from the thermodynamic description of equilibria in macroscopic systems. Such IT investigations have already generated an information semantics of several intuitive chemical concepts and chemical processes; it has also generated an entirely new entropy/information perspective on chemical reactions, e.g., [104–106].

The *open* microscopic systems, however, require the *mixed*-state (*grand*-ensemble), description capable of reflecting the externally imposed thermodynamic conditions, in terms of the *ensemble*-average physical quantities. Since reactivity phenomena involve electronic flows between the *mutually*-open substrates, only in such a generalized framework can one precisely define the relevant reactivity criteria, determine the hypothetical states of subsystems and eventually measure the effects of their mutual interaction. In this ensemble approach, the energetic and *resultant*-information principles are exactly equivalent, giving rise to the same predictions of thermodynamic equilibria. This basic equivalence is consistent with alternative energetic and entropic equilibrium principles invoked in the ordinary thermodynamics of macroscopic systems [81].

1.7 DONOR-ACCEPTOR SYSTEMS

In reactivity considerations one conventionally recognizes several hypo-thetical stages of chemical processes [9, 11, 62–71], involving either the mutually *closed* [nonbonded, disentangled] or *open* [bonded, entangled] reactants $\alpha = \{A, B\}$, e.g., substrates in a typical bimolecular reactive system R = A—B involving the acidic (A, electron acceptor) and basic (B, electron donor) subsystems. The subsystem densities $\{\rho_\alpha = N_\alpha p_\alpha\}$ can be either "frozen," e.g., in the *promolecular* reference $R^0 = (A^0|B^0)$ consisting of the separate (isolated) reactant distributions $\{\rho_\alpha^0 = N_\alpha^0 p^0\}$ shifted to their actual positions in the reactive system, or "polarized" in R^+ = $(A^+|B^+)$, i.e., relaxed in a presence of the reaction partner: $\{\rho_\alpha^+ = N_\alpha^+ p_\alpha^+ = N_\alpha^0 p_\alpha^+\}$. The *nonbonded* status of these fragments, when they conserve their initial overall numbers of electrons of isolated reactants, $\{N_\alpha^+ = N_\alpha^0\}$, is symbolized by the *solid* vertical line separating the two subsystems. It should be emphasized that only due to this mutual closure the substrate identity remains a meaningful concept. In the global equilibrium state of

R as a whole these polarized subsystem density are additionally modified by by the effective *inter*-reactant CT into $\{\rho_\alpha^* = N_\alpha^* p_\alpha^*\}$.

The overall electron density in R^+ as a whole is given by the sum of reactant densities, polarized in the "*molecular*" external potential $v = v_A + v_B$, due to a presence of the reaction partner:

$$\rho_R^+ \equiv N_R p_R^+ = \rho_A^+ + \rho_B^+ \equiv N_A^+ p_A^+ + N_B^+ p_B^+, \, N^+ = \int \rho_\alpha^+ dr = N_\alpha^0, \, \sum_\alpha N_\alpha^+ = N_R. \quad (61)$$

Here, $\{p_\alpha^+ = \rho_\alpha^+/N_\alpha^+\}$ stand for the (internal) probability densities of such promoted fragments, and the system global probability distribution $p_R^+ = \rho_R^+/N_R$ denotes the "shape" function of the overall electron density in the polarized reactive system:

$$p_R^+ = (N_A^+/N_R) p_A^+ + (N_B^+/N_R) p_B^+ \equiv P_A^+ p_A^+ + P_B^+ p_B^+, \, \int p_R^+ dr = P_A^+ + P_B^+ = 1, \quad (62)$$

where the reactant condensed probabilities $\{P_\alpha^+ = N_\alpha^+/N_R = N_\alpha^0/N_R = P_\alpha^0\}$ reflect fragment shares in N_R. At this polarization stage both fragments exhibit internally equalized chemical potentials $\{\mu_\alpha^+ = \mu[N_\alpha^0, v]\}$, modified compared to levels in separate reactants: $\{\mu_\alpha^0 = \mu[N_\alpha^0, v_\alpha] \neq \mu_\alpha^+\}$.

One further observes that the two complementary subsystems lose their identity in their *bonded* status of the *mutually*-open parts of the *externally*-closed reactive system $R \equiv (A^* \vert B^*)$, which allows the *inter*-fragment flow of electrons. The absence of a barrier for such electron transfers has been symbolically represented by the *broken* vertical line separating the two fragments. In such a global equilibrium, marked by the fragment populations for the equalized chemical potentials of both reactants, $\{\mu_\alpha^* = \mu_R[N_R, v]\}$, each "part" effectively explores the probability distribution p_R of the whole reactive system: $\rho_A^* = N_A^* p_R$ and $\rho_B^* = N_B^* p_R$.

However, one can also contemplate, the *external* flows of electrons between the mutually nonbonded reactants and their (separate) external reservoirs $\{\mathcal{R}_\alpha\}$. The mutual closure then implies relevancy of subsystem identities, e.g., that established at the polarization stage in R^+, while the external openness of the composite subsystems $\{\mathcal{M}_\alpha^* = (\alpha^* \vert \mathcal{R}_\alpha)\}$ in

$$\mathcal{M}_R^* = (\mathcal{R}_A \vert A^* \vert B^* \vert \mathcal{R}_B) = (\mathcal{M}_A^* \vert \mathcal{M}_B^*) \quad (63)$$

allows one to independently "regulate" the external chemical potentials of both parts, $\{\mu_\alpha^* = \mu(\mathcal{R}_\alpha)\}$, and hence also their average densities $\{\rho_\alpha^* = N_\alpha^* p_\alpha^*\}$ and electron populations $N_\alpha^* = \int \rho_\alpha^* dr$. In particular, the substrate chemical potentials equalized at the molecular level in both subsystems, $\{\mu_\alpha^* \equiv \mu[N_R, v] = \mu_R\}$, i.e., a common molecular reservoir $\{\mathcal{R}_\alpha(\mu_R) = \mathcal{R}(\mu_R)\}$

coupled to both reactants in $[\mathcal{R}(\mu_R)|\mathrm{A}^*|\mathrm{B}^*]$, formally define also the *global equilibrium* part $\mathrm{R}^* = (\mathrm{A}^*|\mathrm{B}^*) = \mathrm{R}$ of:

$$\mathcal{M}_R^*(\mu_R) \equiv [\mathcal{R}(\mu_R)|\mathrm{A}^*|\mathrm{B}^*] \equiv (\mathcal{R}|\mathrm{A}^*|\mathrm{B}^*) = (\mathcal{R}|\mathrm{R}^*)$$

$$= \mathcal{M}_R^*(\mu_R) \equiv [\mathcal{M}_A^*(\mu_R)|\mathcal{M}_B^*(\mu_R)] = [\mathcal{R}(\mu_R)|\mathrm{A}^*(\mu_R) | \mathrm{B}^+(\mu_R)|\mathcal{R}(\mu_R)]. \qquad (64)$$

This composite reactive system indeed implies an effective *mutual-openness* of both reactants in $\mathcal{M}_R^*$, to a common (molecular) reservoir $\mathcal{R}(\mu_R)$, realized through separate *external*-openness of both substrates in $\mathcal{M}_R^*(\mu_R)$. It allows for an effective donor→acceptor flow of electrons while retaining the fragment identities assured by their *mutual* closeness in $(\mathcal{M}_A^*|\mathcal{M}_B^*)$.

The final, equilibrium reactant distributions $\{\rho_\alpha^* = \rho_\alpha^*(\mu_R)\}$ in R^* and associated electron populations $\{N_\alpha^* = N_\alpha^*(\mu_R) = \int\rho_\alpha^* dr\}$ are modified by the amount of CT:

$$N_{CT} = N_A^* - N_A^0 = N_B^0 - N_B^* > 0, \qquad (65)$$

for the conserved overall number of electrons in the globally isoelectronic processes in the reactive system as a whole:

$$N_A^* + N_B^* \equiv N_R^* = N(\mu_R) = N_R = N_A^+ + N_B^+ \equiv N_R^+ = N_A^0 + N_B^0 \equiv N_R^0. \qquad (66)$$

These *global*-equilibrium redistributions are indexed by the fragment and overall FF in the reactive system. In CSA [68, 69] one introduces the *reactant*-resolved FF matrix of the substrate density responses to the fragment population displacement for the fixed (molecular) external potential:

$$\mathbf{f}^+(\mathbf{r}) = \{f_{\alpha,\beta}(\mathbf{r}) = [\partial\rho_\beta^+(\mathbf{r})/\partial N_\alpha]_v\}, \qquad (67)$$

which generate the *in situ* FF of reactants:

$$f_\alpha^{CT}(\mathbf{r}) = \partial\rho_\alpha^+(\mathbf{r})/\partial N_{CT} = f_{\alpha,\alpha}(\mathbf{r}) - f_{\beta,\alpha}(\mathbf{r}); \ (\alpha, \beta\neq\alpha) \in \{\mathrm{A}, \mathrm{B}\}. \qquad (68)$$

One further recalls, that these relative responses eventually combine into the corresponding global CT derivative, for the reactive system as a whole [9, 11, 69, 72, 73]:

$$F_R^{CT}(\mathbf{r}) = \partial\rho_R^+(\mathbf{r})/\partial N_{CT} = \sum_{\alpha=A,B}\sum_{\beta=A,B} (\partial\rho_\beta^+(\mathbf{r})/\partial N)(\partial N_\alpha/\partial N_{CT})$$

$$= [f_{A,A}(\mathbf{r}) - f_{B,A}(\mathbf{r})] - [f_{B,B}(\mathbf{r}) - f_{A,B}(\mathbf{r})] \equiv f_A^{CT}(\mathbf{r}) - f_B^{CT}(\mathbf{r}), \qquad (69)$$

which represents the populational sensitivity of electron density in $R^* = R$ with respect to the *effective* internal CT between the *externally*-open but *mutually*-closed reactants.

To summarize, the fragment identity remains a meaningful concept only for the *mutually*-closed (nonbonded) status of the acidic and basic reactants, e.g., in the *polarized* reactive system R^+ or in the R^* part of $\mathcal{M}_R^*$. The global equilibrium in R as a whole, $R = R^*$, combining the effectively *"bonded,"* *externally*-open but *mutually*-closed subsystems $\{\alpha^*\}$ in $\mathcal{M}_R^*(\mu_R)$, accounts for the extra CT-induced polarization of reactants compared to R^+. Descriptors of this state, of the mutually "bonded" reactants, can be thus inferred only indirectly, by examining the chemical potential equalization in the composite system $\mathcal{M}_R^*(\mu_R)$. Similar external reservoirs are involved when one examines independent population displacements on reactants, e.g., in defining the fragment chemical potentials and their hardness tensor in R^+ or in the substrate fragment of $\mathcal{M}_R^+$. In this chain of reaction "events" the polarized system R^+ thus appears as the *launching*-stage for the subsequent CT and the accompanying *induced* polarization, after the hypothetical barrier for the flow of electrons between subsystems has been formally lifted.

Thus, the global equilibrium in the macroscopic composite system $\mathcal{M}_R^*(\mu_R) = [\mathcal{M}_A^*(\mu_R)|\mathcal{M}_B^*(\mu_R)] = [\mathcal{R}_A(\mu_R)|A^*(\mu_R)|B^*(\mu_R)|\mathcal{R}_B(\mu_R)]$ also represents the effectively open reactants in $[A^*(\mu_R)|B^*(\mu_R)|\mathcal{R}(\mu_R)] \equiv [R(\mu_R)|\mathcal{R}(\mu_R)]$. These substrate fragments display the equilibrium densities $\{\rho_\alpha^*\}$ after the (implicit) B→A CT, $\rho_A^* + \rho_B^* = \rho_R$ for $\{\rho_\alpha^*/N_\alpha^* = \rho_R/N_R = p_R\}$, and populations $\{\int \rho_\alpha^* dr = N_\alpha^*\}$ corresponding to the chemical potential equalization in $R = R^*$ as a whole:

$$\mu_A^*[\{\rho_\alpha^*\}] = \mu_B^*[\{\rho_\alpha^*\}] = \mu_R[\rho_R]. \tag{70}$$

One recalls, that the reactant chemical potentials have not been equalized at the polarization stage, in the molecular part $R^+ = (A^+|B^+)$ of

$$\mathcal{M}_R^+ = (\mathcal{M}_A^+|\mathcal{M}_B^+) = (\mathcal{R}_A|A^+|B^+|\mathcal{R}_B), \{\mathcal{R}_\alpha = \mathcal{R}_\alpha[\mu_\alpha^+]\}, \tag{71}$$

when $\mu_A^+[\{\rho_\alpha^+\}] < \mu_B^+[\{\rho_\alpha^+\}]$.

The chemical potentials of polarized reactants, $\mu_R^+ = \{\mu_\alpha^+\}$, and elements of the associated hardness matrix $\eta_R^+ = \{\eta_{\alpha\beta}\}$ represent populational derivatives of the system ensemble-average electronic energy in reactant resolution in the *grand*-ensemble representation of $\mathcal{M}_R^+$, $E_v(\{N_\beta\})$, calculated for the fixed external potential v reflecting the "frozen" molecular geometry. These quantities are defined by the corresponding partials of the system *ensemble*-average energy with respect to average electron populations $\{N\}$ on subsystems in the *mutually*-closed (*externally*-open) composite subsystems $\{\mathcal{M}_\alpha^+ = (\alpha^+|\mathcal{R}_\alpha)\}$ of $\mathcal{M}_R^+ = (\mathcal{M}_A^+|\mathcal{M}_B^+)$:

$$\mu_\alpha \equiv \partial E_v(\{N_\gamma\})/\partial N_\alpha, \ \eta_{\alpha,\beta} = \partial^2 E_v(\{N_\gamma\})/\partial N_\alpha \partial N_\beta = \partial\mu_\alpha/\partial N_\beta. \quad (72)$$

The associated global properties of $R = (A^*|B^*) \equiv R^*$ or $\{\mathcal{M}_\alpha^* = (\alpha^*|\mathcal{R})\}$ in $\mathcal{M}_R^*$ similarly involve differentiations with respect to the average number of electrons of R in the combined system $\mathcal{M}_R^* = (R^*|\mathcal{R}) = (A^*|B^*|\mathcal{R})$ $= (\mathcal{R}|A^*|B^*|\mathcal{R})$:

$$\mu_R = \partial E_v(N_R)/\partial N_R, \ \eta_R = \partial^2 E_v(N_R)/\partial N_R^2 = \partial\mu_R/\partial N_R. \quad (73)$$

The optimum amount of the (fractional) CT is then determined by the *in situ* populational "force" measuring the difference between chemical potentials of the polarized acidic and basic reactants in R_n^+, the effective CT gradient,

$$\mu_{CT} = \partial E_v(N_{CT})/\partial N_{CT} = \mu_A^+ - \mu_B^+ < 0, \quad (74)$$

and the *in situ* hardness (η_{CT}) or softness (S_{CT}) for this process [72, 73],

$$\eta_{CT} = \partial\mu_{CT}/\partial N_{CT} = (\eta_{A,A} - \eta_{A,B}) + (\eta_{B,B} - \eta_{B,A}) \equiv \eta_A^R + \eta_B^R = S_{CT}^{-1}, \quad (75)$$

representing the CT-Hessian and its inverse, respectively. The optimum amount of *inter*-reactant CT,

$$N_{CT} = -\mu_{CT} S_{CT} > 0, \quad (76)$$

then generates the associated (2nd-order) stabilization energy:

$$E_{CT} = \mu_{CT} N_{CT}/2 = -\mu_{CT}^2 S_{CT}/2 < 0. \quad (77)$$

It directly follows from the proportionality relations of Eqs. (54) and (58) that the same predictions follow from the associated *in situ* information descriptors.

1.8 HSAB PRINCIPLE

The equivalence of reactivity concepts in the energy and resultant *gradient*-information representations has also direct implications for the Communication Theory of the Chemical Bond (CTCB) [9–12, 23, 24, 30–33]. In its OCT realization, the molecule is regarded as an information channel propagating signals of the Atomic Orbital (AO) origins of electrons in the bond system determined by the occupied MO. The communication noise (orbital indeterminicity) in this network, measured by the channel *conditional*-entropy, is due to electron delocalization in the bond system of a molecule. It represents the system entropic bond "covalency," while the channel *information*-capacity (orbital determinicity), reflected by the *mutual*-information of the molecular communication network, describes the information bond "iconicity." Therefore, the more scattering (indeterminate) is the communication network, the higher its *covalent* character. Accordingly, a less noisy (more deterministic) molecular channel represents more *ionic* information system.

In chemistry the bond covalency, a common possession of electrons by interacting atoms, is indeed synonymous with an electron delocalization generating the communication noise. A classical examples are bonds connecting identical atoms, e.g., hydrogens in H_2 or carbons in ethylene, when the interacting AO in the familiar MO diagrams of chemistry exhibit the same levels of AO energies. The bond ionicity, between atoms exhibiting large differences in their electronegativities, is associated with a substantial CT. Such bonds correspond to a wide separation of the AO energy levels in MO diagrams of chemistry. The ionic bonds introduce more determinicity (less noise) into molecular communications, thus representing the bond mechanism competitive with the bond covalency.

One of the celebrated (qualitative) rules of chemistry deals with the stability preferences in molecular coordination's. The HSAB principle of Pearson [77] predicts that chemically hard (H) acids (A) prefer to coordinate hard bases (B) in the [HH] complexes, and soft (S) acids prefer to coordinate soft bases in [SS] complexes, whereas the "mixed" [HS] or [SH] combinations, of hard acids with soft bases or of soft acids with hard bases, are relatively unstable. This *global* preference is no longer valid *regionally*, between acidic or basic fragments of reactants, where the *complementarity* principle [68, 69, 107, 108] dictates the observed hierarchy of coordination stabilities, conditioned by both covalent and electrostatic interactions [109].

In communication theory of reactive systems [110] the H and S reactants correspond to internally ionic (deterministic) and covalent (noisy) reactant channels, respectively. The former involves *localized* orbital communications between the *mutually*-bonded atoms, while the latter corresponds to strongly *delocalized* information scatterings between AIM. A natural question then arises: what is the preferred overall character of communications corresponding to the mutual interaction between reactants? Do the S substrates of the [SS] complex predominantly interact "covalently," and H substrates in the [HH] complex "ionically"?

In the Frontier Electron (FE) [111–113] approximation of molecular interactions the substrate Highest Occupied MO (HOMO) level determines its *donor* (basic) chemical potential, while the Lowest Unoccupied MO (LUMO) energy establishes its *acceptor* (acidic) capacity [67]. The HOMO-LUMO energy gap then reflects the molecular hardness (Figure 1.2) [44]. One also recalls that the interaction between the reactant MO of comparable energy levels is predominantly *covalent* (chemically "*soft*"), while that between subsystems MO of distinctly different energy levels becomes mostly *ionic* (chemically "*hard*"). A qualitative diagram of Figure 1.2 summarizes the alternative (relative) positions of the donor (HOMO) levels of the basic reactant, relative to the acceptor (LUMO) levels of its acidic partner, for all admissible hardness combinations in the R = A—B coordination system. In view of the proportionality relations of Eqs. (54) and (58), which imply the physical equivalence of the energetic and *gradient*-information reactivity descriptors, these relative MO energy levels also reflect the corresponding information potential and hardness quantities of subsystems, including the *in situ* derivatives driving the CT and information flow between them.

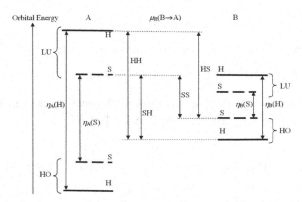

FIGURE 1.2   Schematic diagram of the in situ chemical potentials $\mu_{CT}(B{\to}A)$, determining the effective (internal) CT from basic (B) reactant to its acidic (A) partner in A—B complexes, for their alternative hard (H) and soft (S) combinations. The subsystem hardness descriptors, measured by the HOMO–LUMO gaps in orbital energies, are also indicated.

The magnitude of *ionic* (CT) stabilization energy in A—B systems [Eq. (77)], is determined by the corresponding *in situ* derivatives in R [72],

$$\Delta \varepsilon_{ion.} = |E_{CT}| = \mu_{CT}^{2}/(2\eta_{CT}), \tag{78}$$

where μ_{CT} and η_{CT} stand for the effective CT chemical potential and hardness descriptors of R involving the FE energies of reactants. Since the donor/acceptor properties of the latter are already implied by their *relative* acidic or basic character in R, one applies the *biased* estimate of the CT chemical potential. In this FE approximation the chemical potential difference μ_{CT} for the effective internal B→A CT thus reads (see also Figure 1.2):

$$\mu_{CT}(B \to A) = \mu_{A}^{(-)} - \mu_{B}^{(+)} = \varepsilon_{A}(LUMO) - \varepsilon_{B}(HOMO) \approx I_{B} - A_{A} > 0. \tag{79}$$

It determines the 1ˢᵗ-order energy change for this process:

$$\Delta E_{B \to A}(N_{CT}) = \mu_{CT}(B \to A) N_{CT} > 0. \tag{80}$$

The (positive) *in situ* chemical potential of Eq. (79) combines the electron-*removal* potential of the basic reactant, i.e., its negative ionization potential $I_{B} = E(B^{+1}) - E(B^{0}) > 0$,

$$\mu_{B}^{(+)} = \varepsilon_{B}(HOMO) \approx - I_{B}, \tag{81}$$

and the electron-*insertion* potential of the acidic substrate, i.e., its negative electron affinity $A_{A} = E(A^{0}) - E(A^{-1}) > 0$,

$$\mu_{A}^{(-)} = \varepsilon_{A}(LUMO) \approx -A_{A}, \tag{82}$$

The energy of the CT *disproportionation* process:

$$[A\text{--}B] + [A\text{--}B] \longrightarrow [A^{-1}\text{--}B^{+1}] + [A^{+1}\text{--}B^{-1}], \tag{83}$$

similarly generates the (*unbiased*) *finite*-difference measure of the effective hardness descriptor for this implicit CT [44, 90]:

$$\eta_{CT} = (I_{A} - A_{A}) + (I_{B} - A_{B})$$
$$\approx [\varepsilon_{A}(LUMO) - \varepsilon_{A}(HOMO)] + [\varepsilon_{B}(LUMO) - \varepsilon_{B}(HOMO)]$$
$$= \eta_{A} + \eta_{B} > 0. \tag{84}$$

It ultimately determines the magnitude of CT stabilization energy of Eq. (78), the ionic part of the overall interaction energy:

$$\Delta\varepsilon_{ion.} = \mu_{CT}^2/(2\eta_{CT}) = [\varepsilon_A(LUMO) - \varepsilon_B(HOMO)]^2/[2(\eta_A + \eta_B)]. \tag{85}$$

In the FE framework of Figure 1.2 the CT interaction energy is thus proportional to the squared gap between the LUMO orbital energy of the acidic reactant and the HOMO level of the basic reactant. This ionic interaction is thus predicted to be strongest in the [HH] pair of subsystems and weakest in the SS arrangement, with the mixed [HS] and [SH] combinations representing the intermediate sizes of this *ionic*-stabilization effect.

It should be realized, however, that the ionic and covalent energy contributions compete with one another, complementing each other in the resultant bond energy [67]. Therefore, the [SS]-complex, for which the energy gap $\varepsilon_A(LUMO) - \varepsilon_B(HOMO)$ between the interacting orbital's reaches the minimum value, implies the strongest covalent stabilization of the reaction complex. Indeed, the lowest (bonding) MO energy level ε_b in this FE interaction, corresponding to the bonding combination of the interacting orbital's of subsystems:

$$\varphi_b = N_b[\varphi_B(HOMO) + \lambda\varphi_A(LUMO)], \tag{86}$$

then exhibits the maximum *covalent* effect:

$$\Delta\varepsilon_{cov.} = \varepsilon_B(HOMO) - \varepsilon_b > 0.$$

It is limited by the inequality:

$$\Delta\varepsilon_{cov.} < (\beta - \varepsilon_b S)^2/[\varepsilon_A(LUMO) - \varepsilon_B(HOMO)], \tag{87}$$

where the MO-coupling matrix element of the system electronic Hamiltonian H,

$$\beta = \langle\varphi_A(LUMO)|H|\varphi_B(HOMO)\rangle, \tag{88}$$

is expected to be proportional to the orbital overlap integral $S = \langle\varphi_A(LUMO)|\varphi_B(HOMO)\rangle$.

It thus follows from Eq. (87) that the maximum covalent component of the *inter*-reactant chemical bond is expected in interactions between soft (strongly overlapping) reactants, since then the numerator assumes the highest value while the denominator reaches its minimum. For the same reason one predicts the smallest covalent stabilization in interactions between the (weakly overlapping) hard reactants, with the mixed hardness combinations giving rise to intermediate levels of the bond covalency.

To summarize, the globally hard complex of the [HH]-coordination exhibits the maximum *ionic*-stabilization, the globally soft [SS]-complex corresponds to the maximum *covalent*-interaction, while the mixed combinations of reactant hardnesses in [HS]- and [SH]-pairs exhibit a mixture of moderate covalent and ionic bonds between acidic and basic subsystems. Therefore, electron communications representing the *inter*-reactant bonds between the chemically soft (covalent) reactants are also expected to be predominantly "soft" (delocalized, indeterministic) in character, while those between the chemically hard (ionic) subsystems are predicted to be dominated by the "hard" (localized, deterministic) propagations in the communication system for R as a whole.

The electron communications between reactants α = A, B in the *donor-acceptor* coordination system R = AB are determined by the corresponding matrix of conditional probabilities in AO resolution or of their amplitudes, which can be partitioned into the corresponding *intra*-reactant parts, combining internal communications within individual substrates, and the *inter*-reactant blocks of external communications between different subsystems:

$$[R{\to}R] = \{[\alpha{\to}\beta]\} = \{[\alpha{\to}]\delta_{\alpha,\beta}\} + \{[\alpha\beta]\,(1 - \delta_{\alpha,\beta})\} = \{intra\} + \{inter\}. \qquad (89)$$

Therefore, in most stable complexes the "soft" (noisy), delocalized internal blocks of such probability propagations imply similar covalent character of the external blocks of communications between reactants, i.e., strongly indeterministic scatterings between subsystems, $\{intra\text{-}S\} \Rightarrow \{inter\text{-}S\}$. The "hard" (ionic), internal channels are similarly associated with the ionic (localized) external communications: $\{intra\text{-}H\} \Rightarrow \{inter\text{-}H\}$. This observation adds a new, *communication*-angle to the classical HSAB principle of chemistry.

1.9   CONCLUSION

The applicability of IT in explaining diverse issues in theory of electronic structure of molecules and their preferences in chemical reactions have already been demonstrated elsewhere [9–12,68–74,114–120]. In this overview we have examined the QIT description of the bimolecular donor-acceptor reactive systems, including all hypothetical processes that accompany bond-breaking–bond-forming processes in chemical reactions.

Continuities of the classical (modulus/probability) and nonclassical (phase/current) state parameters have been summarized, contributions these degrees-of-freedom generate in the resultant entropy or information measures have been identified, and implications of the probability and phase continuities for the net production of the overall gradient information have been examined. In this perspective we have argued that QIT facilitates a deeper understanding of general rules of chemistry and provides a "thermodynamical" description of the *externally*-open molecular systems and their fragments. The proportionality relation between the resultant gradient-information and kinetic energy allows one to apply the virial theorem partitioning in explaining general rules of chemical reactivity.

The need for nonclassical (phase/current) supplements of the classical (probability) measures of the entropy/information content in molecular states has been reemphasized. To paraphrase Prigogine [82], the electron density distribution determines only a "static" facet of molecular electronic structure, which we call the structure of "being," while the current distribution describes its "dynamic" aspect, called the structure of "becoming." Both these manifestations of a molecular "organization" contribute to the overall information content in generally complex electronic states, as explicitly reflected by the resultant QIT concepts combining the classical (probability) and nonclassical (phase/current) components of the state overall entropy/information content. Their importance in describing the mutual bonding and nonbonding status of molecular fragments, for the same set of electron densities in subsystems, has been stressed and the *in situ* populational derivatives of the *ensemble*-average values of the energy and *gradient*-information functionals have been explored as alternative reactivity indicators.

The DFT-based theory of chemical reactivity distinguishes several hypothetical stages involving either the mutually bonded (entangled) or nonbonded (disentangled) reactants, for the same set of electron distributions in constituent subsystems. These two categories are discerned only by the *phase*-aspect of the quantum entanglement of such molecular fragments. The equilibrium phases and currents of reactants can be related to electron densities using the entropic principles of QIT. This generalized approach deepens our understanding of promotions in molecular fragments and provides a more precise framework for monitoring the reaction progress and its intermediate stages vital for understanding in chemistry.

In QIT the populational derivatives of the resultant *gradient*-information, proportional to the system average kinetic energy, have been suggested as alternative reactivity criteria, equivalent to familiar energetical descriptors. They were shown to correctly predict both the direction and magnitude of the electron flows and associated information propagations in reactive systems. The *grand*-ensemble description of thermodynamic equilibria in the *externally* open molecular systems has been outlined and the physical equivalence of variational principles for electronic energy and resultant gradient information has been stressed. The virial theorem partitioning of energy profiles has been applied to explain the qualitative Hammond postulate of reactivity theory and the information production in chemical reactions has been addressed. The ionic and covalent interactions between frontier electrons of the acidic and basic reactants have been examined to justify the HSAB principle of chemistry and to provide the communication outlook on the interaction/communication between acids and bases. It has been argued that the *internally* soft and hard reactants prefer to *externally* communicate in the like manner, consistent with their internal communication pattern. This preference should be also reflected by the *inter*-reactant bonds: covalent in [SS]-complex and ionic in [HH]-coordination.

KEYWORDS

- **equilibrium principles**
- **grand-ensemble**
- **HSAB principle**
- **reactivity criteria**
- **resultant information**
- **virial partitioning**

REFERENCES

1. Fisher, R. A., (1925). Theory of statistical estimation. *Proc Cambridge Phil. Soc.,* *22*, 700–725.
2. Frieden, B. R., (2004). *Physics from the Fisher Information: A Unification.* Cambridge University Press, Cambridge.

3. Shannon, C. E., (1948). The mathematical theory of communication. *Bell System Tech. J., 27,* 379–493, 623–656.
4. Shannon, C. E., & Weaver, W., (1949). *The Mathematical Theory of Communication.* University of Illinois, Urbana.
5. Kullback, S., & Leibler, R. A., (1951). On information and sufficiency. *Ann. Math Stat., 22,* 79–86.
6. Kullback, S., (1959). *Information Theory and Statistics.* Wiley, New York.
7. Abramson, N., (1963). *Information Theory and Coding.* McGraw-Hill, New York.
8. Pfeifer, P. E., (1978). *Concepts of Probability Theory.* Dover, New York.
9. Nalewajski, R. F., (2006). *Information Theory of Molecular Systems.* Elsevier, Amsterdam.
10. Nalewajski, R. F., (2010). *Information Origins of the Chemical Bond.* Nova Science Publishers, New York.
11. Nalewajski, R. F., (2012). *Perspectives in Electronic Structure Theory.* Springer, Heidelberg.
12. Nalewajski, R. F., (2016). *Quantum Information Theory of Molecular States.* Nova Science Publishers, New York.
13. Nalewajski, R. F., & Parr, R. G., (2000). Information theory, atoms-in-molecules, and molecular similarity. *Proc. Natl. Acad. Sci. USA, 97,* 8879–8882.
14. Nalewajski, R. F., (2003). Information principles in the theory of electronic structure. *Chem. Phys. Lett., 272,* 28–34.
15. Nalewajski, R. F., (2003). Information principles in the loge theory. *Chem. Phys. Lett., 375,* 196–203.
16. Nalewajski, R. F., & Broniatowska, E., (2003). Information distance approach to Hammond postulate. *Chem. Phys. Lett., 376,* 33–39.
17. Nalewajski, R. F., & Parr, R. G., (2001). Information-theoretic thermodynamics of molecules and their Hirshfield fragments. *J. Phys. Chem. A, 105,* 7391–7400.
18. Nalewajski, R. F., (2002). Hirshfeld analysis of molecular densities: Subsystem probabilities and charge sensitivities. *Phys. Chem. Chem. Phys., 4,* 1710–1721.
19. Parr, R. G., Ayers, P. W., & Nalewajski, R. F., (2005). What is an atom in a molecule? *J. Phys. Chem. A, 109,* 3957–3959.
20. Nalewajski, R. F., & Broniatowska, E., (2007). Atoms-in-molecules from the stockholder partition of molecular two-electron distribution. *Theoret. Chem. Acc., 117,* 7–27.
21. Heidar-Zadeh, F., Ayers, P. W., Verstraelen, T., Vinogradov, I., Vöhringer-Martinez, E., & Bultinck, P., (2018). Information-theoretic approaches to atoms-in-molecules: Hirshfeld family of partitioning schemes. *J. Phys. Chem. A, 122,* 4219–4245.
22. Hirshfeld, F. L., (1977). Bonded-atom fragments for describing molecular charge densities. *Theoret. Chim. Acta (Berl), 44,* 129–138.
23. Nalewajski, R. F., (2000). Entropic measures of bond multiplicity from the information theory. *J Phys. Chem. A, 104,* 11940–11951.
24. Nalewajski, R. F., (2004). Entropy descriptors of the chemical bond in information theory: I. Basic concepts and relations. *Mol. Phys., 102,* 531–546; II. Application to simple orbital models. *Mol. Phys., 102,* 547–566.
25. Nalewajski, R. F., (2004). Entropic and difference bond multiplicities from the two-electron probabilities in orbital resolution. *Chem. Phys. Lett., 386,* 265–271.

26. Nalewajski, R. F., (2005). Reduced communication channels of molecular fragments and their entropy/information bond indices. *Theoret. Chem. Acc., 114*, 4–18.
27. Nalewajski, R. F., (2005). Partial communication channels of molecular fragments and their entropy/information indices. *Mol. Phys., 103*, 451–470.
28. Nalewajski, R. F., (2011). Entropy/information descriptors of the chemical bond revisited. *J. Math. Chem., 49*, 2308–2329.
29. Nalewajski, R. F., (2014). Quantum information descriptors and communications in molecules. *J. Math. Chem., 52*, 1292–1323.
30. Nalewajski, R. F., (2009). Multiple, localized, and delocalized/conjugated bonds in the orbital-communication theory of molecular systems. *Adv. Quant. Chem., 56*, 217–250.
31. Nalewajski, R. F., Szczepanik, D., & Mrozek, J., (2011). Bond differentiation and orbital decoupling in the orbital communication theory of the chemical bond. *Adv. Quant. Chem., 61*, 1–48.
32. Nalewajski, R. F., Szczepanik, D., & Mrozek, J., (2012). Basis set dependence of molecular information channels and their entropic bond descriptors. *J. Math. Chem., 50*, 1437–1457.
33. Nalewajski, R. F., (2017). Electron communications and chemical bonds. In: Wójcik, M., Nakatsuji, H., Kirtman, B., & Ozaki, Y., (eds.), *Frontiers of Quantum Chemistry* (pp. 315–351). Springer, Singapore.
34. Nalewajski, R. F., Świtka, E., & Michalak, A., (2002). Information distance analysis of molecular electron densities. *Int. J. Quantum. Chem., 87*, 198–213.
35. Nalewajski, R. F., & Broniatowska, E., (2003). Entropy displacement analysis of electron distributions in molecules and their Hirshfeld atoms. *J. Phys. Chem. A, 107*, 6270–6280.
36. Nalewajski, R. F., (2008). Use of Fisher information in quantum chemistry. *Int. J. Quantum Chem.* (Jankowski, K., issue), *108*, 2230–2252.
37. Nalewajski, R. F., Köster, A. M., & Escalante, S., (2005). Electron localization function as information measure. *J. Phys. Chem. A, 109*, 10038–10043.
38. Becke, A. D., & Edgecombe, K. E., (1990). A simple measure of electron localization in atomic and molecular systems. *J. Chem. Phys., 92*, 5397–5403.
39. Silvi, B., & Savin, A., (1994). Classification of chemical bonds based on topological analysis of electron localization functions. *Nature, 371*, 683–686.
40. Savin, A., Nesper, R., Wengert, S., & Fässler, T. F., (1997). ELF: The electron localization function. *Angew. Chem. Int. Ed. Engl., 36*, 1808–1832.
41. Hohenberg, P., & Kohn, W., (1964). Inhomogeneous electron gas. *Phys. Rev., 136B*, 864–971.
42. Kohn, W., & Sham, L. J., (1965). Self-consistent equations including exchange and correlation effects. *Phys. Rev., 140A*, 1133–1138.
43. Levy, M., (1979). Universal variational functionals of electron densities, first-order density matrices, and natural spin-orbital's and solution of the *v*-representability problem. *Proc. Natl. Acad. Sci. USA, 76*, 6062–6065.
44. Parr, R. G., & Yang, W., (1989). *Density-Functional Theory of Atoms and Molecules*. Oxford University Press, New York.
45. Dreizler, R. M., & Gross, E. K. U., (1990). *Density Functional Theory: An Approach to the Quantum Many-Body Problem*. Springer, Berlin.

46. Nalewajski, R. F., (1996). Density functional theory I-IV. *Topics in Current Chemistry, 180–183.*
47. Nalewajski, R. F., De Silva, P., & Mrozek, J., (2010). Use of nonadditive Fisher information in probing the chemical bonds. *J. Mol. Struct: THEOCHEM, 954*, 57–74.
48. Nalewajski, R. F., (2011). Through-space and through-bridge components of chemical bonds. *J. Math. Chem., 49*, 371–392.
49. Nalewajski, R. F., (2011). Chemical bonds from through-bridge orbital communications in prototype molecular systems. *J. Math. Chem., 49*, 546–561.
50. Nalewajski, R. F., (2011). On interference of orbital communications in molecular systems. *J. Math. Chem., 49*, 806–815.
51. Nalewajski, R. F., & Gurdek, P., (2011). On the implicit bond-dependency origins of bridge interactions. *J. Math. Chem., 49*, 1226–1237.
52. Nalewajski, R. F., (2012). Direct (through-space) and indirect (through-bridge) components of molecular bond multiplicities. *Int. J. Quantum Chem., 112*, 2355–2370.
53. Nalewajski, R. F., & Gurdek, P., (2012). Bond-order and entropic probes of the chemical bonds. *Struct. Chem., 23*, 1383–1398.
54. Nalewajski, R. F., (2016). Complex entropy and resultant information measures. *J. Math. Chem., 54*, 1777–1782.
55. Nalewajski, R. F., (2014). On phase/current components of entropy/information descriptors of molecular states. *Mol. Phys., 112*, 2587–2601.
56. Nalewajski, R. F., (2017). Quantum information measures and their use in chemistry. *Current Phys. Chem., 7*, 94–117.
57. Nalewajski, R. F., (2013). Exploring molecular equilibria using quantum information measures. *Ann. Phys. (Leipzig), 525*, 256–268.
58. Nalewajski, R. F., (2014). On phase equilibria in molecules. *J. Math. Chem., 52*, 588–612.
59. Nalewajski, R. F., (2014). Quantum information approach to electronic equilibria: Molecular fragments and elements of non-equilibrium thermodynamic description. *J. Math. Chem., 52*, 1921–1948.
60. Nalewajski, R. F., (2015). Phase/current information descriptors and equilibrium states in molecules. *Int. J. Quantum Chem., 115*, 1274–1288.
61. Nalewajski, R. F., (2015). Quantum information measures and molecular phase equilibria. In: Baswell, A. R., (ed.), *Advances in Mathematics Research* (Vol. 19, pp. 53–86). Nova Science Publishers New York.
62. Nalewajski, R. F., (2018). Phase description of reactive systems. In: Islam, N., & Kaya, S., (eds.), *Conceptual Density Functional Theory* (pp. 217–249). Apple Academic Press, Waretown.
63. Nalewajski, R. F., (2017). Entropy continuity, electron diffusion and fragment entanglement in equilibrium states. In: Baswell, A. R., (ed.), *Advances in Mathematics Research* (Vol. 22, pp. 1–42). Nova Science Publishers, New York.
64. Nalewajski, R. F., (2016). On entangled states of molecular fragments. *Trends in Physical Chemistry, 16*, 71–85.
65. Nalewajski, R. F., (2017). Chemical reactivity description in density-functional and information theories. In: Liu, S., (ed.), Chemical concepts from density functional theory. *Acta Physico-Chimica Sinica, 33*, 2491–2509.

66. Nalewajski, R. F., (2018). Information equilibria, subsystem entanglement, and dynamics of overall entropic descriptors of molecular electronic structure. *J. Mol. Model.* (Chattaraj, P. K., issue), *24*, 212–227.

67. Nalewajski, R. F., (2019). On entropy/information description of reactivity phenomena. In: Baswell, A. R., (ed.), *Advances in Mathematics Research* (Vol. 26, pp. 97–157). Nova Science Publishers, New York.

68. Nalewajski, R. F., Korchowiec, J., & Michalak, A., (1996). Reactivity criteria in charge sensitivity analysis. In: Nalewajski, R. F., (ed.), *Topics in Current Chemistry: Density Functional Theory IV* (Vol. 183, pp. 25–141). Springer, Berlin.

69. Nalewajski, R. F., & Korchowiec, J., (1997). *Charge Sensitivity Approach to Electronic Structure and Chemical Reactivity.* World Scientific. Singapore.

70. Geerlings, P., De Proft, F., & Langenaeker, W., (2003). Conceptual density functional theory. *Chem. Rev., 103*, 1793–1873.

71. Chattaraj, P. K., (2009). *Chemical Reactivity Theory: A Density Functional View.* CRC Press, Boca Raton.

72. Nalewajski, R. F., (1994). Sensitivity analysis of charge transfer systems: *In situ* quantities, intersecting state model and its implications. *Int. J. Quantum Chem., 49*, 675–703.

73. Nalewajski, R. F., (1995). Charge sensitivity analysis as diagnostic tool for predicting trends in chemical reactivity. In: Dreizler, R. M., & Gross, E. K. U., (eds.), *Proceedings of the NATO ASI on Density Functional Theory (Il Ciocco, 1993)* (pp. 339–389). Plenum, New York.

74. Gatti, C., & Macchi, P., (2012). *Modern Charge-Density Analysis.* Springer, Berlin.

75. Nalewajski, R. F., (1980). Virial theorem implications for the minimum energy reaction paths. *Chem. Phys., 50*, 127–136.

76. Hammond, G. S., (1955). A correlation of reaction rates. *J. Am. Chem. Soc., 77*, 334–338.

77. Pearson, R. G., (1973). *Hard and Soft Acids and Bases.* Dowden, Hutchinson, and Ross, Stroudsburg.

78. Weizsäcker, C. F. von, (1935). Zur theorie der kernmassen. *Z. Phys., 96*, 431–458.

79. Harriman, J. E., (1980). Orthonormal orbitals for the representation of an arbitrary density. *Phys. Rev. A, 24*, 680–682.

80. Zumbach, G., & Maschke, K., (1983). New approach to the calculation of density functionals. *Phys. Rev. A, 28*, 544–554; *Erratum, Phys. Rev. A*, 29, 1585–1587.

81. Callen, H. B., (1962). *Thermodynamics: An Introduction to the Physical Theories of Equilibrium Thermostatics and Irreversible Thermodynamics.* Wiley, New York.

82. Prigogine, I., (1980). *From Being to Becoming: Time and Complexity in the Physical Sciences.* Freeman, W. H., & Co, San Francisco.

83. Leeuwen, R. van, Gritsenko, O. V., & Baerends, E. J., (1996). Analysis and modelling of atomic and molecular potentials. In: Nalewajski, R. F., (ed.), *Topics in Current Chemistry: Density Functional Theory I* (Vol. 180, pp. 107–167). Springer, Berlin.

84. Marcus, R. A., (1968). Theoretical relations among rate constants, barriers, and Broensted slopes of chemical reactions. *J. Phys. Chem., 72*, 891–899.

85. Agmon, N., & Levine, R. D., (1977). Energy, entropy, and the reaction coordinate: Thermodynamic-like relations in chemical kinetics. *Chem. Phys. Lett., 52*, 197–201.

86. Agmon, N., & Levine, R. D., (1979). Empirical triatomic potential energy surfaces defined over orthogonal bond-order coordinates. *J. Chem. Phys., 71,* 3034–3041.
87. Miller, A. R., (1978). A theoretical relation for the position of the energy barrier between initial and final states of chemical reactions. *J. Am. Chem. Soc., 100,* 1984–1992.
88. Ciosłowski, J., (1991). Quantifying the Hammond postulate: Intramolecular proton transfer in substituted hydrogen catecholate anions. *J. Am. Chem. Soc., 113,* 6756–6761.
89. Nalewajski, R. F., Formosinho, S. J., Varandas, A. J. C., & Mrozek, J., (1994). Quantum mechanical valence study of a bond-breaking–bond-forming process in triatomic systems. *Int. J. Quantum Chem., 52,* 1153–1176.
90. Nalewajski, R. F., & Broniatowska, E., (2003). Information distance approach to Hammond postulate. *Chem. Phys. Lett., 376,* 33–39.
91. Dunning, T. H., Jr., (1984). Theoretical studies of the energetics of the abstraction and exchange reactions in H + HX, with X = F–I. *J. Phys. Chem., 88,* 2469–2477.
92. Ruedenberg, K., (1962). The physical nature of the chemical bond. *Rev. Mod. Phys., 34,* 326–376.
93. Feinberg, M. J., & Ruedenberg, K., (1971). Paradoxical role of the kinetic-energy operator in the formation of the covalent bond. *J. Chem. Phys., 54,* 1495–1512.
94. Feinberg, M. J., & Ruedenberg, K., (1971). Heteropolar one-electron bond. *J. Chem. Phys., 55,* 5805–5818.
95. Bacskay, G. B., Nordholm, S., & Ruedenberg, K., (2018). The virial theorem and covalent bonding. *J. Phys. Chem. A, 122,* 7880–7893.
96. Gyftopoulos, E. P., & Hatsopoulos, G. N., (1965). Quantum-thermodynamic definition of electronegativity. *Proc. Natl. Acad. Sci., USA, 60,* 786–793.
97. Perdew, J. P., Parr, R. G., Levy, M., & Balduz, J. L., (1982). Density functional theory for fractional particle number: Derivative discontinuities of the energy. *Phys. Rev. Lett., 49,* 1691–1694.
98. Mulliken, R. S., (1934). A new electronegativity scale: Together with data on valence states and on ionization potentials and electron affinities. *J. Chem. Phys., 2,* 782–793.
99. Iczkowski, R. P., & Margrave, J. L., (1961). Electronegativity. *J. Am. Chem. Soc., 83,* 3547–3551.
100. Parr, R. G., Donnelly, R. A., Levy, M., & Palke, W. E., (1978). Electronegativity: The density functional viewpoint. *J. Chem. Phys., 69,* 4431–4439.
101. Parr, R. G., & Pearson, R. G., (1983). Absolute hardness: Companion parameter to absolute electronegativity. *J. Am. Chem. Soc., 105,* 7512–7516.
102. Parr, R. G., & Yang, W., (1984). Density functional approach to the frontier-electron theory of chemical reactivity. *J. Am. Chem. Soc., 106,* 4049–4050.
103. Neumann, N. J., von, (1955). *Mathematical Foundations of Quantum Mechanics.* Princeton University Press, Princeton.
104. Hô, M., Schmider, H. L., Weaver, D. F., Smith, Jr. V. H., Sagar, R. P., & Esquivel, R. O., (2000). Shannon entropy of chemical changes: S_N2 displacement reactions. *Int. J. Quantum Chem., 77,* 376–382.
105. López-Rosa, S., Esquivel, R. O., Angulo, J. C., Antolin, J., Dehesa, J. S., & Flores-Gallegos, N., (2010). Fisher information study in position and momentum spaces for elementary chemical reactions. *J. Chem. Theory Comput., 6,* 145–154.

106. Esquivel, R. O., Liu, S. B., Angulo, J. C., Dehesa, J. S., Antolin, J., & Molina-Espiritu, M., (2011). Fisher information and steric effect: Study of the internal rotation barrier in ethane. *J. Phys. Chem. A, 115*, 4406–4415.

107. Nalewajski, R. F., (2000). Manifestations of the maximum complementarity principle for matching atomic softnesses in model chemisorption systems. *Topics in Catalysis, 11*, 469–485.

108. Chandra, A. K., Michalak, A., Nguyen, M. T., & Nalewajski, R. F., (1998). On regional matching of atomic softnesses in chemical reactions: Two-reactant charge sensitivity study. *J. Phys. Chem. A, 102*, 10182–10188.

109. Nalewajski, R. F., (1984). Electrostatic effects in interactions between hard (soft) acids and bases. *J. Am. Chem. Soc., 106*, 944–945.

110. Nalewajski, R. F., (2014). Quantum information description of reactive systems. *Indian J. Chem.* (Ghosh, S. K., issue), *53A*, 1010–1018.

111. Fukui, K., (1975). *Theory of Orientation and Stereoselection.* Springer-Verlag, Berlin.

112. Fukui, K., (1987). Role of frontier orbitals in chemical reactions. *Science, 218*, 747–754.

113. Fujimoto, H., & Fukui, K., (1974). Intermolecular interactions and chemical reactivity. In: Klopman, G., (ed.), *Chemical Reactivity and Reaction Paths* (pp. 23–54.). Wiley-Interscience, New York.

114. Nalewajski, R. F., (2018). Information equilibria, subsystem entanglement and dynamics of overall entropic descriptors of molecular electronic structure. *J. Mol. Model.* (Chattaraj, P. K., issue), *24*, 212–227.

115. Nalewajski, R. F., (2019). Understanding electronic structure and chemical reactivity: Quantum-information perspective. In: Sousa, S., (ed.), *The Application of Quantum Mechanics to the Reactivity of Molecules. Appl. Sci., 9*, 1262–1292.

116. Nalewajski, R. F., (2020). Phase-equalization, information flows and electron communications in donor-acceptor systems. In: Sousa, S., (ed.), *The Application of Quantum Mechanics to the Reactivity of Molecules. Appl. Sci., 10*, 3615–3646.

117. Nalewajski, R. F., (2020). Information-theoretic descriptors of molecular states and electronic communications between reactants. In: Matta, C. F., (ed.), *Information Theoretic Approaches to Atoms and Molecules. Entropy, 22*, 749–769.

118. Nalewajski, R. F., (2020). Role of electronic kinetic energy and resultant gradient information in chemical reactivity. In: Berski, S., & Sokalski, W. A., (eds.), *J. Mol. Model.* (Latajka, Z., issue), *25*, 259–278.

119. Nalewajski, R. F., (2019). Equidensity orbitals in resultant-information description of electronic states. *Theoret. Chem. Acc., 138*, 108–123.

120. Nalewajski, R. F., (2019). Resultant information description of electronic states and chemical processes. *J. Phys. Chem. A* (Geerlings, P., issue), *123*, 45–60.

CHAPTER 2

A Computational Modeling of the Structure, Frontier Molecular Orbital (FMO) Analysis, and Global and Local Reactive Descriptors of a Phytochemical 'Coumestrol'

P. VINDUJA, VIJISHA K. RAJAN, SWATHI KRISHNA, and K. MURALEEDHARAN

Department of Chemistry, University of Calicut, Malappuram – 673635, Kerala, India

ABSTRACT

Coumestrol is an estrogenic isoflavonoid which belongs to the class of phytochemicals known as coumestans, with interesting therapeutic applications such as antioxidant and anticancer properties. Coumestrol is widely spread in leguminous plants, alfalfa, ladino clover, strawberry, soya bean, sprouts, pea silage, and beans. The structure of coumestrol is closely resembles to the (*E*)-4,4'-dihydroxystilbene derivatives, which have promising pharmacological activity, and is responsible for the estrogenic activity of coumestrol. The present work, deals with the computational investigation of the structural analysis of coumestrol, by using M06 as level of theory and 6–31++G (d, p) as basis set in Gaussian 09 software package. The stable conformer of coumestrol has been identified through the potential energy scan (PES) and the lowest energy conformer is selected for further investigation. Before going to the deep knowledge of bioactivities shown by the title compound, one must have in-depth knowledge about the molecular structure of the compound. The most robust computational tool, density functional theory (DFT) has been enhanced to a great extent, especially for the structural analysis of organic compounds, a computational exploration into the structural analysis of the title compound is particularly relevant. Detailed structural analysis has been done using H^1 and C^{13} NMR as well as UV-visible spectroscopy. Moreover, the reports showing the detailed structural characterization of coumestrol is found to be rare and the reported papers are mostly focused on its bioactivities. In this scenario, the present work aims to explore a detailed computational approach towards the structural characterization of coumestrol. The work clearly explains the molecular structure, spectral characterizations, and frontier molecular orbital analysis. In addition, the global reactive and Fukui indices of the title compound have also been explained. The work can be extended to detailed bioactivity analysis, molecular docking analysis, QSAR/QSPR studies, etc.

2.1 INTRODUCTION

Computational chemistry is a technique used for the investigation of chemical processes by using a computer. The applications of computers in chemistry are manifold. It is mainly used in the expanse of molecular geometry, energies of molecules and transition state, chemical reactivity, IR, UV, and nuclear magnetic resonance (NMR) spectra, interaction of substrate with an enzyme, physical properties of the substance, multipoles, polarizability, atomic charges, electrostatic potential, magnetic properties, vibrational frequencies, etc. One may think for the necessity of clubbing an experiment with theory. Computational/theoretical methods can be used for interpretation of ambiguous or conflicting results. If there requires a real-time analysis of the progress of experimental process, computational method can be used for optimizing the data. Not all process can be experimentally carried forward, but with computational methods, it can be made easier.

Material science as well as other fields of chemistry is mostly reliable upon theoretical chemistry. More precisely, we can say that computational chemistry is the quantum mechanical study of fundamentals of atoms, molecules, and chemical reactions using principles of mathematical algorithms, statistics, and large databases to integrate chemical theory and impersonate with experimental observations. The present study deals with the structural characterization of an isoflavone, Coumestrol by using NMR as well as UV-Vis Spectroscopy. The medicinal activities demonstrated isoflavones attract excellent attention in studies and their structural characterization plays a vital role in improving their knowledge of bioactivities. The NMR spectroscopy is unique in its ability to characterize a molecule properly and has been an indispensable technique of characterization in both biological and pharmaceutical chemistry [1]. UV spectroscopy is used to identify their energy levels, transition probabilities, and other characteristics. This work presents a computational approach to the NMR and UV-Vis spectroscopic analysis of a potentially active isoflavone Coumestrol in a detailed manner.

Structurally similar groups like isoflavones, isoflavanones, isoflavans, pterocarpans, rotenoids come under the class of isoflavanoids [2]. Coumestrol is an important member of the family of natural compounds namely phytoestrogens [3]. Phytoestrogens are the compounds with estrogenic activity which are derived from plants which consist of compounds such as Isoflavones (genistein and diadzein), Coumestans (Coumestrol), lignans (enterolactone and enterodiol), and stilbenes (resveratrol). Coumestans are the phytochemical from where the Coumestrol can be derived. Coumestrol is mostly found in leguminous plants such as Alfalfas [4], ladino clover [5, 6] Chinese herb such as Radix Puerariae and also in plants such as Soybean [7], Brussels, sprouts, spinach, peas, and beans [8], strawberry [9], etc. It has same structure as that of isoflavones and estradiol. Based on the estrogen levels in the body, it shows different estrogenic and anti-estrogenic activities. Estrogenic activity of Coumestrol is due to its stilbene like structure analogs to that of diethylstilbestrol [10].

Recently, people are very much interested in understanding the nature of bioactivities of Coumestans. Coumestrol comes under the category Coumastans whose core structure is closely similar to Coumestan, [1] Benzoxolo [3,2-c] chromen-6-one (Figure 2.1).

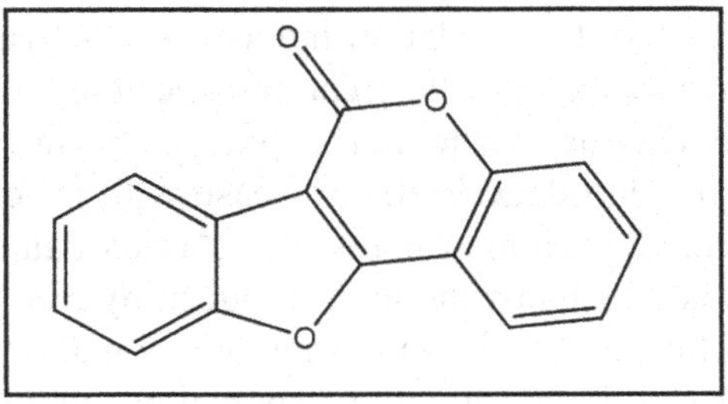

FIGURE 2.1   Basic structure of coumestan.

In the case of Coumestrol, the structure is modified by attaching two hydroxyl groups at 3rd and 9th positions as 3,9-dihydroxy-[1]benzofuro[3,2-c]chromen-6-one (Figure 2.2).

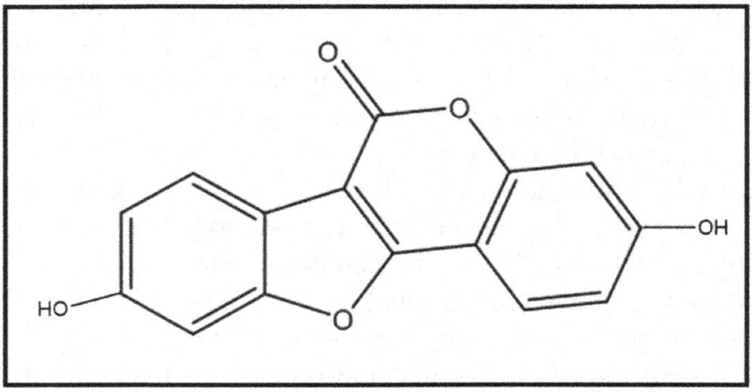

FIGURE 2.2   Basic structure of coumestrol.

It was Bickoff et al. who reported several researches on Coumestrol and accentuated its pharmacological effects [11]. Oxygen being the ultimate electron acceptor for producing energy in the form of ATP in electron flow systems, oxidation responses is an indispensable component of aerobic life. But when uncoupled, the electron stream system produces free radicals, mostly oxygenated species called Reactive Oxygen Species, and these extremely reactive species, including radicals of superoxide, peroxy radicals, alkoxy radicals, etc., cause damage to life operations. Oxidation is liable for pathogenesis of multiple age-related degenerative diseases such as cancer, diabetes, muscular degeneration, Alzheimer's and Parkinson's disease, since reactive pro-oxidant species can harm proteins, lipids, carbohydrates, and nucleic acids over time [12]. Furthermore, several studies have suggested that oxidation induces and activates multiple cell signaling pathways that add to the creation of lesions of toxic substances and ultimately potentiate Alzheimer's disease. Alzheimer's disease is the most common Neurodegenerative disease with progressive memory, cognition, and behavioral impairments [13]. Antioxidants are compounds that safeguard the cell against the harm induced by free radical chemical reactions. The inclusion of fruits and vegetables containing antioxidants in the diet is very crucial in this situation. They either stabilize or deactivate the free radicals before them attacking cells. Antioxidants are the first line of protection against reactive oxygen species and are critical to keeping optimum activity in the cell and life. Here, Coumestrol which is a potent anti-oxidant shows anti-Alzheimer's property. Cerebral ischemia is a condition in which there is insufficient blood flow to the brain to meet metabolic requirements but even in this case Coumestrol because of its low molecular weight and stable structure, enables it to transit through the cell membrane, and shows its neuroprotective nature [14]. Intake of food-stuffs containing Coumestrol has an irrefutable effect in cancer treatment, menopause difficulties, problems related to osteoporosis, atherosclerosis, and cardiovascular disease [15]. Reports have been shown that Coumestrol has a positive influence in anti-aging [16]. Adipogenesis is a process of conversion of pre-adipocytes into adipocytes which eventually causes the excessive fat deposition. In this scenario, Coumestrol acts as a powerful tool for anti-adipogenesis. Melanin is the pigment that gives color to skin and hair and it absorbs UV light and protects the skin against the harmful impact of UV light and free radicals generated by UV. Despite the beneficial role of melanin in the skin, excessive melanin production and accumulation outcomes in several skin disorders including acquired syndromes of hyperpigmentation such as melasma, age spots, and freckles. In this case, coumestrol acts as a depigmenting agent. Our studies described here were designed to assess the antioxidant activity of Coumestrol.

2.2 MATERIALS AND METHODOLOGY

2.2.1 MATERIALS

The present study deals with the computational investigation of structural analysis of coumestrol. The input structures are taken from Pubchem software which is in SDF format and is converted into GJF using Openbabel application. Some modifications on the structure are done using the Gauss-view-5.0 graphical user interface and all the computational parameters are calculated with the help of the Gaussian 09 software package.

2.2.2 COMPUTATIONAL METHODOLOGY

The present work is purely based on the density functional theory (DFT) analysis of structure of a Coumestan, coumestrol. Calculations are performed on the molecule by setting M06 as level of theory and 6–31++G(d, p) as a basis set. Computational investigations are performed using Gaussian 09 software.

In order to get the stable conformer of coumestrol potential energy scan (PES) is done. Among all the possible conformers, one with lowest energy is selected for future studies. The selected conformer is optimized by choosing 631++G (d, p) basis set.

2.2.3   FRONTIER MOLECULAR ORBITAL ANALYSIS

Frontier molecular orbital (FMO) Analysis makes a sense that HOMO and LUMO orbitals are actively participating in chemical reaction. It is the redistribution of electrons results in chemical bonding. HOMO is the highest occupied molecular orbital which energetically favor the removal of electrons from it and acts as a Lewis base while LUMO is the lowest unoccupied orbital which energetically favors the acceptance of electron and acts as a Lewis acid [11]. The HOMO-LUMO gap or called bandgap of coumestrol is calculated by analyzing the HOMO and LUMO orbitals of optimized coumestrol. The bandgap of the molecule will give a clear idea about the reactivity of the molecule. Lowest band gap indicates that the molecule is highly polarized and shows high reactivity [12]. From the bandgap, it is possible to study the charge transfer interactions between donor and acceptor moiety. Chemical stability can be explained with the help of hardness and softness of the molecule. Higher the bandgap hard will be the molecule and the lower bandgap is a clear indication of soft molecule. Soft molecules are very easy to polarize than hard molecules which are very difficult to polarize [13].

Three-dimensional charge distributions of charges on the molecule is obtained from electrostatic potential maps so that the way by which molecules are interacting each other can be studied and that again leads to the information regarding the charge related properties of molecule. Electrostatic potential maps give the strength of nearby charges, nuclei, molecules, and will give a clear idea about the behavior of complexes.

2.2.4 GLOBAL REACTIVE DESCRIPTIVE PARAMETERS

The relation between the chemical reactivity of a molecule and its sensitivity towards structural perturbations and external conditions can be described using global descriptive parameters [17]. Assigning a single value to the entire molecule is possible through a global descriptive tool. These tools are used for describing the intrinsic reactivity of the molecule as a whole. The first ionization potential, IP, and electron affinity, EA, of the system are the simplest global reactivity descriptors. These are used for measuring the tendency of the system to donate or accept one electron. For a system with N_0 electrons, these are defined as:

$$IP = E\,(N_0-1) - E\,(N_0) \tag{1}$$

$$EA = E\,(N_0) - E\,(N_0+1) \tag{2}$$

There are two methods for calculating global reactive descriptors; one is known as 'energy vertical,' which is obtained from the geometry of the neutral molecule and is the difference between the total electronic energy of the neutral molecule to that of its corresponding cation and anion. It is given by the following equation:

$$\text{Vertical ionization potential (VIP)} = E_{cation} - E_{neutral} \tag{3}$$

$$\text{Vertical electron affinity (VEA)} = E_{neutral} - E_{anion} \tag{4}$$

Another method is, by using Koopmann's theorem;

$$\text{Ionization potential (IP)} \approx -E_{HOMO} \tag{5}$$

$$\text{Electron affinity (EA)} \approx -E_{LUMO} \tag{6}$$

Where E_{HOMO} is the energy of HOMO and E_{LUMO} is the energy of LUMO.

Furthermore, we know the electron-donating power is denoted by $\omega-$, and the electron-accepting power is denoted by ω^+, which are defined in terms of the square of the chemical potential, $(\mu^\pm)^2$, and by the common hardness or resistance, η, associated with both processes and is given by the equation:

$$\omega^\mp = (\mu^\mp)^2/2\eta \tag{7}$$

This shows the tendency of the system to donate or accept a charge in fractional quantities without saturation. It also depends on the exchange of electronic charge with the surrounding environment. Usually, chemical hardness defines the unwillingness of an atom, ion, or molecule to undergo a deformation or polarization under small perturbation experienced during a chemical reaction, whereas chemical softness is defined as the capacity of a molecule to accept electrons. It is usually converse of chemical hardness. Chemical potential is the measure of tendency of an electron to escape from equilibrium and chemical potential in DFT is explained by the first derivative of energy with respect to the number of electrons. Here further studies show that ω^+ call for a greater electron acceptance power and this charge acceptance power stabilizes the system, whereas ω^- demands for greater electron-donating capacity; thereby destabilizing the system.

$$\omega^- = (3IP + EA)^2 / 16 (IP - EA) \tag{8}$$

$$\omega^+ = (IP + 3EA)^2 / 16 (IP - EA) \tag{9}$$

In these equations, up to this approximation, we can see that the electron-donating power, which is a reactive index in which IP outweighs EA but whereas the electron-accepting power shows the opposite character. So finally, the electrophilicity relationship between global indices of reactivity gives the same value to IP as well as EA without distinguishing between the charge-donating and charge-accepting processes. This index appraises only the stabilization in energy when the system acquires the maximum amount of electronic charge from its surroundings. Not only this, Here:

$$\mu \approx -(IP + EA)/2 \text{ and} \tag{10}$$

$$\eta \approx (IP - EA)/2 \tag{11}$$

The electrophilicity is used for the description of reactivity that allows a quantitative classification of the global electrophilic nature of a molecule within a relative scale. Electrons flow from places with higher chemical potential to places with lower chemical potential until the chemical potential of both systems are equal. This gives the electronegativity as the negative of the chemical potential:

$$\chi_e = -\mu. \tag{12}$$

If $\in_{HOMO}$ and $\in_{LUMO}$ are the energies of the highest occupied and lowest unoccupied molecular orbital's (LUMOs), respectively, then the above equations can be rewritten as:

$$\chi = -(\epsilon_{HOMO} + \epsilon_{LUMO})/2 \text{ and} \qquad (13)$$

$$\eta = (\epsilon_{LUMO} - \epsilon_{HOMO})/2 \qquad (14)$$

There are two classifications for global reactivity indicators from notional DFT; fundamental and derived. For global reactivity tools, ChemTools support the built-in and user-defined energy models listed below. Using the code of these models as templates, users can develop new energy models in ChemTools; Linear Energy Model, Quadratic Energy Model, Exponential Energy Model, Rational Energy Model, General Energy Model.

2.2.5 TIME-DEPENDENT DENSITY FUNCTIONAL THEORY (TDDFT)

TDDFT which is an elegant extension of DFT for complex, dynamic excited state systems [18]. TDDFT is now commonly used to investigate the excitation spectrum of extended solids and molecules [19–21] due to its relatively low computational cost compared to wave function and practical Green's function-based approaches. DFT and TDDFT are same in principle, but in practice, precision is constrained by the approximations currently available for the exchange-correlation (xc) contribution to the total energy functional E_{xc}, and its derivative interaction kernel (by second functional derivatives), f_{xc} [22]. TDDFT includes methods of measuring electronic excitation energies formally precise and technically convenient. TDDFT incorporates some common DFT ideas, most notably the Kohn-Sham principle of replacing the actual multi-body interacting system with a non-interacting system reproducing the same density. But there are also many terms, such as memory and initial-state dependency that are unique to the time-dependent case [23].

2.3 RESULTS AND DISCUSSION

2.3.1 OPTIMIZATION

The potential energy surface scan of Coumestrol has been performed and the lowest energy conformer with energy –5.7905 eV has been identified.

3-D PES image of Coumestrol is given in Figure 2.3 and lowest energy-optimized structure of Coumestrol is given in Figure 2.4.

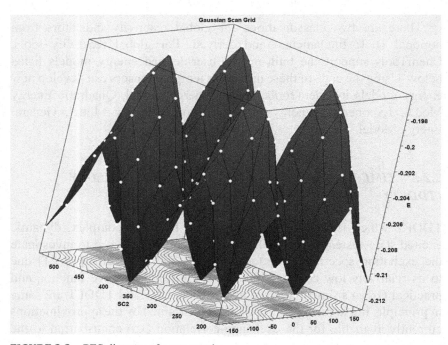

FIGURE 2.3   PES diagram of coumestrol.

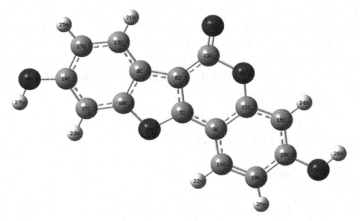

FIGURE 2.4   Optimized coumestrol molecule.

2.3.2 GLOBAL REACTIVE DESCRIPTIVE PARAMETERS

To predict the chemical reactivity both global and local descriptive parameters are widely used. Chemical hardness, chemical softness, chemical potential, electronegativity, electrophilicity, ionization energy, EA, etc., comes under global descriptive parameters. These properties are essential for analyzing the chemical nature of a compound. Here in the case of Coumestrol, a comparative study between the vertical energy calculation and Koopmans's theorem has been performed and results are illustrated in Table 2.1. The term electronegativity implies the ability of an atom to pull electrons towards it whereas chemical potential is defined as the tendency of an electron to move across a potential gradient until it becomes uniform. The stability of the molecule is projected through its hardness. Chemical hardness basically means the unwillingness of the electron cloud of atoms, ions, or molecules to undergo deformation for minute disturbance during chemical procedures and it directly linked to HOMO-LUMO energy gap of the molecule. The larger IP obtained in both the methods, i.e., vertical energy method (7.42 eV) and Koopmann's method (6.14 eV) shows the intricacy to take away electrons from Coumestrol. The capacity of an atom to receive electrons is termed as chemical softness. Chemical hardness is inversely proportional to chemical softness. The measure of escaping tendency of an electron from equilibrium is generally termed as chemical potential but in DFT it is defined as the first derivative of energy with respect to number of electrons. Strength of electrophilicity of a species is obtained from its electrophilicity index. The lower EA value obtained in the case of Coumestrol reflects its higher electron attracting capacity.

TABLE 2.1 Global Descriptive Parameters (eV) of Coumestrol

Method	Ionization Potential (IP)	Electron Affinity (EA)	Electro-negativity (χ)	Hardness (η)	Softness (s)	Chemical Potential (μ)	Electro-philic Index (ω)
Vertical energy calculation	7.42	0.46	3.94	3.47	0.14	−3.94	2.23
Koopmann's	6.14	1.73	3.93	2.2	0.227	−3.93	3.51

2.3.3   NMR SPECTROSCOPIC ANALYSIS

Detailed structural analysis of the isoflavanone Coumestrol is analyzed by means of NMR technique using M06/631++g(d,p) computational methodology. The computed H^1-NMR and C^{13}-NMR spectra of Coumestrol are shown in Figures 2.5 and 2.6, respectively.

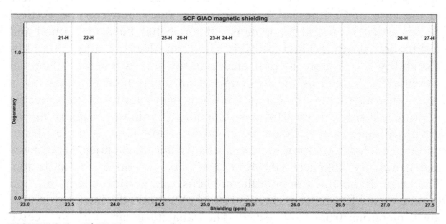

FIGURE 2.5 H^1 NMR spectra of coumestrol.

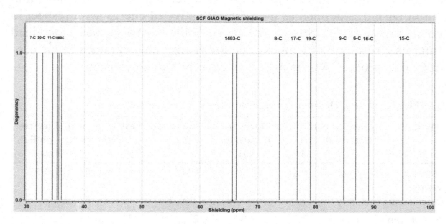

FIGURE 2.6 C^{13} NMR spectra of coumestrol.

2.3.3.1 H^1 NMR SPECTRAL ANALYSIS

The H^1-NMR spectrum of Coumetrol clearly indicates that there are 8 different protons existing at different chemical environments. All the protons are chemically non-equivalent. The aromatic protons resonate nearly in the region 6–8 ppm and in coumestrol, also these are observed in the region between 6.2–7.7 ppm. All the hydroxylic protons resonate at 4–5 ppm. H25, H21, H23, H24, H26, H22, H24 are part of sp^2 hybridized carbon atoms of aromatic rings. Aromatic protons usually resonate nearly in the region 6–8 ppm and here in Coumestrol, the protons H23, H24, H25, H26, resonates at 6.87, 6.78, 7.46 and 7.27 eV, respectively. But an abnormality from the regular trend is observed for the aromatic protons H21 and H22, which resonates at 8.55 and 8.27 eV, respectively. Transfer of electron density from the hydrogen atom to an electronegative atom is the common way of hydrogen bonding interaction and this will result in the deshielding of proton which in turn increase the chemical shift value. Stronger the hydrogen bonding greater will be the chemical shift value. The higher chemical shift value of H21 is attributed to its attachment to the carbon nearest to carbonyl oxygen atom. So in the case of H21 proton there is a possibility of intra molecular hydrogen bonding interaction hence higher chemical shift. In the same way, H23 proton also experiences a higher chemical shift. It can also because of the same reason; intramolecular hydrogen bonding with neighboring oxygen atom present in the heterocyclic ring system. An electronegative atom present in the heterocyclic ring system will attract electron density from the alkyl group near to it, hence increases the deshielding which eventually causes the increase in chemical shift. It is also observed that all the hydroxylic protons will resonate in the region between 4–5 ppm. A great correlation found between the principles of chemical shifts in NMR spectroscopy and computed H^1-NMR spectrum of Coumestrol.

2.3.3.2 C^{13} NMR SPECTRAL ANALYSIS

C^{13}-NMR of Coumestrol reveals that there are 15 different carbon atoms. The resonating range of aromatic carbon atom is 110–140 ppm. But in this case, the observed range is much higher than the expected one. The reason may be delineated as; the repulsive interaction between electron clouds of protons and the hindering groups present in the rigid molecules promotes the deshielding of these protons. There is a considerable shift in δ value in a positive direction for those C=C of alkenyl carbons which are attachment to more electronegative atoms/ groups. It is C18 and C20 which bear electronegative OH group and shows higher δ value 165.78 and 169.08 ppm, respectively. Normally carbon atoms in the aromatic ring show a general chemical shift value in the range 125–170. Here in the case of carbon atoms C17, C13, C8, C10, C15 of aromatic ring near to the five-membered heterocyclic rings shows chemical shifts 125.09, 135.62, 128.19, 166.53 and 106.89 ppm. It is clear that C15 shows an abnormal behavior from normal trend. Aromatic carbons on the ester ring C6, C7, C9, C11 shows chemical shift values 115.01, 170.06, 117.09, and 167.36 ppm, respectively. The aromatic carbon atoms C16, C19, C14 resonate at 112.76, 122.85, and 136.29 ppm, respectively.

2.3.4   UV-VISIBLE SPECTROSCOPY

The electronic transitions in a molecule can be easily interpreted by means of its UV-Visible spectrum. In computational methodology, the UV-Visible spectral analysis can be done by a TDDFT analysis under same level of theory and basis set as that in optimization studies. The experimental and computed UV-Visible spectra of Coumestrol are given in Figure 2.7a–2.7d.

Coumestans usually exhibit absorption peaks in the region 340 to 355 nm and 230–250 nm. There are two prominent peaks observed in the gas phase absorption spectra, in which λ_{max} has been observed at 331.43 nm and 240.2 nm. In gas-phase maximum absorption (331.43 nm) is due to the HOMO to LUMO transition and the peak at 240.2 nm corresponds to the transition from HOMO-4 to LUMO. Other low-intensity peaks are observed at 283.58, 258.14, 253.74, 216.24, and 210.54 nm which is due to the transitions HOMO to LUMO+4, HOMO-2 to LUMO, HOMO-1 to LUMO, HOMO-1 to LUMO+3, and HOMO-2 to LUMO+3, respectively. The computed UV-Visible spectral characteristics are given in Table 2.2.

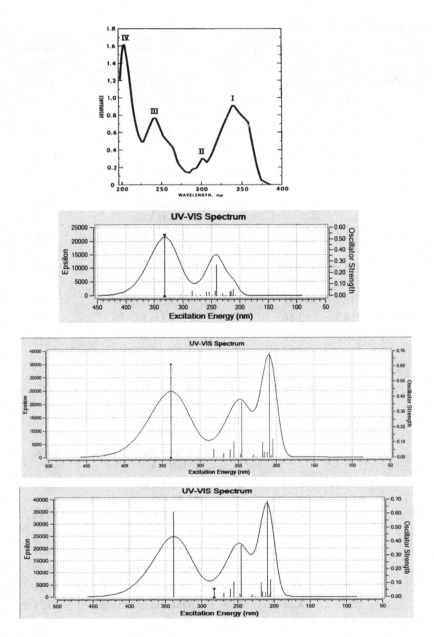

FIGURE 2.7 a) Experimental UV-visible spectrum of coumestrol in methanol, computed UV-Visible spectra of Coumestrol in b) gas phase, c) methanol and d) water.

TABLE 2.2   UV Spectral Parameters of Coumestrol

Medium	λ_{max} (nm)	Energy (ev)	Oscillator Strength	Orbital's Involved	% Contribution of MO
Gas Phase	331.43	3.74	0.53	HOMO to LUMO	97.03
	240.2	5.16	0.01	HOMO-4 to LUMO	76.71
	283.58	4.37	0.04	HOMO to LUMO+4	76.42
	258.14	4.80	0.03	HOMO-2 to LUMO	23.64
	253.74	4.88	0.03	HOMO-1 to LUMO	46.23
	216.24	5.73	0.01	HOMO-1 to LUMO+3	42.49
	210.54	5.88	0.01	HOMO-2 to LUMO+3	47.49
	211.45	5.86	0.05	HOMO-2 to LUMO+4	29.98
	216.90	5.72	0.03	HOMO to LUMO+14	49.92
	242.70	5.11	0.04	HOMO-3 to LUMO	59.90
Methanol	338.57	3.66	0.61	HOMO to LUMO	97.69
	208.93	5.93	0.69	HOMO-3 to LUMO+3	28.30
	245.35	5.05	0.36	HOMO-2 to LUMO	26.23
	204.45	6.06	0.12	HOMO-1 to LUMO+6	41.03
	206.06	6.02	0.01	HOMO −3 to LUMO +2	22.68
	212.01	5.82	0.03	HOMO-2 to LUMO+3	33.96
	216.01	5.74	0.03	HOMO-1 to LUMO+3	28.08
	218.02	5.68	0.09	HOMO-2 to LUMO+3	39.39
	255.72	4.84	0.10	HOMO-1 to LUMO	45.64
	260.54	4.76	0.05	HOMO-2 to LUMO	38.98
	269.04	4.61	0.03	HOMO to LUMO+6	48.83
	282.3	4.39	0.05	HOMO to LUMO+3	70.51
Water	338.66	3.66	0.61	HOMO to LUMO	97.69
	245.43	5.05	0.37	HOMO-2 to LUMO	26.72

TABLE 2.2   *(Continued)*

Medium	λ_{max} (nm)	Energy (ev)	Oscillator Strength	Orbital's Involved	% Contribution of MO
	208.98	5.93	0.69	HOMO-3 to LUMO+3	27.92
	204.53	6.06	0.12	HOMO-1 to LUMO+6	41.01
	212.04	5.85	0.03	HOMO-2 to LUMO+3	33.70
	216.03	5.74	0.03	HOMO-1 to LUMO+3	27.76
	218.06	5.68	0.09	HOMO-1 to LUMO+3	39.99
	246.86	5.02	0.02	HOMO-3 to LUMO	47.09
	255.78	4.85	0.11	HOMO-1 to LUMO	45.59
	260.61	4.75	0.05	HOMO-2 to LUMO	38.74
	269.04	4.61	0.03	HOMO to LUMO+6	48.79
	282.28	4.39	0.05	HOMO to LUMO+3	70.30

UV spectrum of coumestrol in methanol showed three prominent peaks at 208.93, 245.35, and 338.57 nm. The transition from HOMO to LUMO results in the most intense peak at 338.57 nm. The other two sharp peaks are due to HOMO-3 to LUMO+3 and HOMO-2 to LUMO transitions. In addition, some low intensity peaks are observed at 204.45, 206.06, 212.01, 216.01, 218.02, 255.72, 260.54, 269.04, and 282.3 nm. Three intense peaks at 338.66, 245.43, and 208.98 nm have been observed in the UV spectra of coumestrol in water.

2.4   CONCLUSION

The structural analysis of Coumestrol has been computed successfully. Computational investigation of the structural analysis of coumestrol was performed by using M06 as a level of theory and 6–31++G (d, p) as basis set in Gaussian 09 software package. The stable conformer of coumestrol with energy-5.79057 eV has been identified through the PES. Further investigation of the lowest energy conformer has been established. The HOMO-LUMO energy gap and global descriptive parameters showed it has high EA. Detailed structural analysis has been done using H^1 and C^{13} NMR as well as UV-Visible spectroscopy. All the chemical shift values which have been computed through Gaussian 09 software are in good agreement with the principles of NMR spectroscopy. Results obtained in UV-Vis spectroscopy is in good agreement with the experimental results. The computed molecular structure, spectral characterizations, and FMO analysis give a vivid picture about the title compound; Coumestrol. The work can be extended to detailed bioactivity analysis, molecular docking analysis, QSAR/QSPR studies, etc.

ACKNOWLEDGMENTS

The authors are thankful to the Central Sophisticated Instrumentation Facility (CSIF) for providing Gaussian 09 software support. The authors are also thankful to all the colleagues in the Computational Lab, Department of Chemistry, University of Calicut, Malappuram, Kerala.

KEYWORDS

- **coumestrol**
- **density functional theory**
- **global descriptive parameters**
- **isoflavonoids**
- **nuclear magnetic resonance spectroscopy**
- **UV-visible spectroscopy**

REFERENCES

1. Kaminski, P., & Katz, R., (2015). *Yerba Santa Eriodictyon Californicum.* Flower Essence Soc.
2. Birt, D. F., Hendrich, S., & Wang, W., (2001). Dietary agents in cancer prevention: Flavonoids and isoflavonoids. *Pharmacol. Ther., 90*(2/3), 157–177.
3. Bedell, S., Nachtigall, M., & Naftolin, F., (2014). *J. Steroid. Biochem. Mol. Biol., 139*, 225–236.

4. Kirihata, Y., Kawarabayashi, T., Imanishi, S., Sugimoto, M., & Kume, S., (2008). Coumestrol decreases intestinal alkaline phosphatase activity in post-delivery mice but does not affect vitamin D receptor and calcium channels in post-delivery and neonatal mice. *J. Reprod. Dev., 54*(1), 35–41.

5. Livingston, L. A., Hendrickson, A. P., & Booth, A. N., (1962). Relative potencies of several estrogen-like compounds found in forages. *Jour. Agr. Food Chern., 10*(5), 410–412.6.

6. Guggolz, J., Livingston, A. L., & Bickoff, E. M., (1961). Detection of daidzein, formononetin, genistein, and biochanin a in forages. *Jour. Agr. Food Chern., 9*, 330–332.

7. Wada, H., & Yuhara, M., (1964). Identification of plant estrogens in Chinese milk vetch, soybean, and soybean sprout. *Jap. Jour. Zootech. Sci., 35*, 87–91.

8. Bickoff, E. M., Spencek, R. R., Witt, S. C., & Knuckles, B. E., (1958). New estrogen synthesized. *Chern. and Engin. News, 36*(38), 49–50.

9. Morley, F. W. H., & Axelsen, A., (1968). Bioassay responses of ewes to legume swards. ii. uterine weight results from swards. *Austral. Jour. Agr. Res., 18*, 495–504.

10. Biggers, J. D., (1959). Plant phenols possessing oestrogenic activity. In: Fairbairn, J. W., (ed.), *The Pharmacology of Plant Phenolics* (pp. 51–69). Academic Press, London.

11. Bickoff, E. M., Spencerm, R. R., Witt, S. C., & Knuckles, B. E., (1969). *Studies on the Chemical and Biological Properties of Coumestrol and Related Compounds* (Vol. 1408, pp. 1–83). US Department of Agriculture: Washington, DC, USA.

12. Thapa, A., & Carroll, N. J., (2017). Dietary modulation of oxidative stress in Alzheimer's disease. *Int. J. Mol. Sci., 18*, 1583.

13. Rajan, V. K., & Muraleedharan, K., (2017). A computational investigation on the structure, global parameters and antioxidant capacity of a polyphenol, Gallic acid. *Food Chemistry, 220*, 93–99.

14. Castro, C. C., Pagnussat, A. S., Orlandi, L., Worm, P., Moura, N., Etgen, A. M., & Netto, C. A., (2012). Coumestrol has neuroprotective effects before and after global cerebral ischemia in female rats. *Brain Res., 1474*, 82–90.

15. Humfrey, C. D., (1998). Phytoestrogens and human health effects: Weighing up the current evidence. *Nat. Toxins, 6*, 51–59.

16. Park, G., Baek, S., Kim, J. E., Lim, T. G., Lee, C. C., Yang, H., Kang, Y. G., Park, J. S., Augustin, M., Mrosek, M., et al., (2015). Flt3 is a target of coumestrol in protecting against UVB-induced skin photoaging. *Biochem. Pharmacol., 98*, 473–483.

17. Vijisha, K. R., Hasna, C. K., & Muraleedharan, K., (2018). The natural food colorant Peonidin from cranberries as a potential radical scavenger-A DFT based mechanistic analysis. *Food Chemistry, 262*, 184–190.

18. Runge, E., & Gross, E. K. U., (1984). Density-functional theory for time-dependent systems. *Phys. Rev. Lett., 52*, 997.

19. Petersilka, M., Gossmann, U. J., & Gross, E. K. U., (1996). Excitation energies from time-dependent density-functional theory. *Phys. Rev. Lett., 76*, 1212.

20. Bauernschmitt, R., Häser, M., Treutler, O., & Ahlrichs, R., (1997). Calculation of excitation energies within time-dependent density functional theory using auxiliary basis set expansion. *Chem. Phys. Lett., 264*, 573.

21. Stratmann, R. E., Scuseria, G. E., & Frisch, M. J., (1998). *J. Chem. Phys., 109*, 8218.
22. Okan, K. O., & David, D. O., (2019). TDDFT+U: A critical assessment of the Hubbard U correction to exchange-correlation kernels and potentials. *Phy. Rev. B., 99*, 165120.

CHAPTER 3

Theoretical Analysis of CuTiS$_2$ and CuTiSe$_2$ Invoking Density Functional Theory-Based Descriptors

PRABHAT RANJAN,[1] PANCHAM KUMAR,[2] and
TANMOY CHAKRABORTY[3]

[1]*Department of Mechatronics Engineering,
Manipal University Jaipur, Dehmi Kalan, Jaipur – 303007, India,
E-mail: prabhat23887@gmail.com*

[2]*School of Electrical Skills, Bhartiya Skill Development University,
Jaipur – 302042, India*

[3]*Department of Chemistry, School of Engineering,
Presidency University, Bengaluru – 560064, India,
E-mail: tanmoychem@gmail.com*

ABSTRACT

In this report, we have systematically investigated the physicochemical properties of semiconducting materials CuTiS$_2$ and CuTiSe$_2$ in terms of density functional theory-based descriptors. The study of CuTiS$_2$ and CuTiSe$_2$ materials are of importance due to its potential applications in solar cells, light-emitting diodes (LEDs), and nonlinear optical devices. Density functional theory (DFT) is one of the most successful techniques in computational material science and engineering to compute the stability, structure, and electronic, optical, and magnetic properties of materials. Geometry optimization is done with exchange-correlation functional local spin density approximation (LSDA) and basis set Lanl2DZ. The DFT-based descriptors namely, highest occupied molecular orbital (HOMO)-lowest unoccupied molecular orbital (LUMO), electronegativity, hardness, softness, electrophilicity index, and dipole moment are computed. A close agreement between experimental and computed bond lengths is observed from this analysis.

3.1   INTRODUCTION

Renewable energy sources, especially solar energy is gaining a lot of interest among researchers in recent times. Solar energy is known for zero-emission, noise-free, and safe source of energy. The energy radiated from the sun on earth surface is so enormous that if utilized properly, the energy demand of a year globally can meet the requirement from a one-hour radiation from the sun. There are a number of articles in which properties, applications, and change in the efficiency of solar cells is well documented [1–6]. In recent years, third-generation solar cells are becoming popular, as it provides much better efficiency as compared to traditional solar cells. In the case of traditional solar cells, electrons move from valence band to conduction band, whereas in third-generation solar cells, an intermediate band is placed in between the valence and conduction band in such a way so that electrons either can be transferred directly from valence band to the conduction band or can be routed from valence band to the intermediate band first and then intermediate band to the conduction band [7]. In this process, more than one intermediate band can be placed to achieve more efficiency. The efficiency for third-generation solar cells are found as 63% and can raise up to 86% by introducing more number of intermediate bands, which is much higher as compared to single-junction solar cells [1, 2, 7–9]. It is already reported that the optical absorption can be enhanced by introducing an intermediate band in the host materials [10–12].

The study of copper-based chalcopyrite-type materials have been proven effective for light-emitting diodes (LEDs), nonlinear optical devices, and photovoltaic's applications with an energy conversion efficiency of 46.7% [13–16]. The chalcopyrite materials such as $CuGaS_2$ and $CuAlSe_2$ are found suitable for photovoltaic applications due to the high energy gap. It is reported that in transition metal-doped semiconductor material $CuGaS_2$, the optical absorption and photovoltaic properties are enhanced after replacing Ga atom with Ti atom [5, 17]. Transition metals such as titanium, vanadium, and iron-doped $CuAlSe_2$ is reported in which titanium doped $CuAlSe_2$ is the best suitable material for photovoltaic applications [7]. The physicochemical properties of $CuGaS_2$ can be altered by the addition of either donor Ti or acceptor Fe atom [6]. The study shows that $M_xA_2\text{-}xB_2X_4$ (A= Ga or Cu; X= S and Se and M= C, Si, Ge, Sn, V, Ir, Fe, Co, Ni, Rh, and Hg) materials have applications in photovoltaics and spintronics devices [13]. The compound $CuTi_2S_4$ is of importance because it is a donor molecular system which has metallic character and Pauli paramagnetism behavior [18]. The study found that for compound $CuAlX_2$ (X= S, Se, and Te), the calculated energy gap is maximum for $CuAlS_2$ and least gap is obtained for material $CuAlTe_2$ [19]. It shows that energy gap decreases when compound moves from $CuTiS_2$ to $CuTiTe_2$. The compound $CuTiSe_2$ have high adsorption coefficient in the visible range. Recently, we have also studied chalcopyrite type semiconductor materials invoking density functional theory (DFT) methodology. DFT methodology has been proven very successful technique in computing the physicochemical properties, stability, and geometry of metallic doped compounds [20–28]. In this report, we have studied $CuTiS_2$ and $CuTiSe_2$ compounds by using DFT-based descriptors. The DFT-based descriptors such as highest occupied molecular orbital (HOMO)-lowest unoccupied molecular orbital (LUMO), molecular hardness, softness, electronegativity; electrophilicity index and dipole moment are computed.

3.2 COMPUTATIONAL DETAILS

Since last couple of years, DFT has been dominant and effective computational technique for metallic doped semiconducting materials. DFT methodologies are open to various domains of material science, physics, chemistry, surface science, nanoscience, biology, life sciences, pharmacy, and earth sciences [49–51]. In this chapter, we have investigated the computational analysis of $CuTiS_2$ and $CuTiSe_2$ by using DFT technique. Material modeling and geometry optimization have been done using Gaussian 03 [52] within DFT framework. For geometry optimization, local spin density approximation (LSDA) exchange correlation with basis set LANL2DZ is used.

Invoking Koopmans' approximation [49], we have calculated ionization energy (I) and electron affinity (A) of all the compounds using the following approach:

$$I = -\varepsilon_{HOMO} \tag{1}$$

$$A = -\varepsilon_{LUMO} \tag{2}$$

Thereafter, using I and A, the conceptual DFT-based descriptors viz. electronegativity (χ), global hardness (η), molecular softness (S) and electrophilicity index (ω) have been computed. The equations used for such calculations are as follows:

$$\chi = -\mu = \frac{I+A}{2} \tag{3}$$

where, μ represents the chemical potential of the system.

$$\eta = \frac{I-A}{2} \tag{4}$$

$$S = \frac{1}{2\eta} \tag{5}$$

$$\omega = \frac{\mu^2}{2\eta} \tag{6}$$

3.3   RESULTS AND DISCUSSION

The physical and chemical properties of compounds $CuTiS_2$ and $CuTiSe_2$ have been performed invoking DFT technique. The charge transfer in materials is related to the donor-acceptor interaction which eventually governs by the symmetry of the frontier molecular orbital's and change in the electronegativity of the interacting fragments [53–56]. The frontier orbital's of HOMO and LUMO of both donor and acceptor are responsible for charge transfer and bonding [54, 55]. The contribution from the donor HOMO into the acceptor LUMO and then afterwards contribution from the HOMO of the acceptor into the LUMO of the donor is a concurrent process which occurs during the establishment of donor-acceptor in the materials [53–56]. The electronegativity equalization principle is qualitatively associated with charge transfer between the donor and acceptor, it indicates when two reactants come closer, and electrons drift from lower electronegativity to higher electronegativity until it becomes equal at some intermediate value [57–62]. The chemical potential equalization also works on the concept of charge transfer, which state that the charge transfer happens from higher to lower chemical potential system [63–69]. The orbital energies in form of HOMO–LUMO energy gap and DFT-based descriptors such as molecular hardness, softness, electronegativity, and electrophilicity index is presented in Table 3.1. The HOMO-LUMO energy gap is an important factor to study the electronic properties of materials. It shows the minimum amount of energy needed for an electron to move from valence band to conduction band [70]. The HOMO-LUMO energy gap is also associated with the reactivity of compounds. The compound with a minimum value of energy gap is highly reactive and less stable whereas, compound with high value of energy gap are more stable and least reactive. The result from Table 3.1 shows that CuTiSe$_2$ have high energy gap of 0.471 eV whereas, CuTiS$_2$ have energy gap of 0.463 eV. It indicates that CuTiSe$_2$ is more stable as compared to the compound CuTiS$_2$. The result also displays that compound having highest HOMO-LUMO energy gap have maximum value of molecular hardness, electronegativity, and electrophilicity index. The compound CuTiSe$_2$ have maximum value of electronegativity, electrophilicity index, and hardness which are 5.038 eV, 53.918 eV and 0.235 eV respectively. Similarly, HOMO-LUMO energy gap is having inverse relation with molecular softness values. The compound CuTiS$_2$ have high value of softness, i.e., 2.162 eV whereas CuTiSe$_2$ have least value, i.e., 2.124 eV.

The Quadrupole moment according to the Buckingham convention for compound CuTiS$_2$ and CuTiSe$_2$ is reported in Table 3.2. The quadrupole moment values in different axes are presented in Debye-Ang.

A comparative analysis between experimental and theoretical bond lengths for compounds CuTiS$_2$ and CuTiSe$_2$ is shown in the Table 3.3. The computed bond length for compound CuTiS$_2$ are d(Ti-S) = 2.683 Å, d(Cu-S) = 2.256 Å and d(Cu-Ti) = 2.506 Å, which are in agreement with the experimental bond lengths [71]. For compound CuTiSe$_2$, the bond lengths are d(Cu-Se) = 2.375 Å, d(Se-Se) = 2.538 Å, d(Ti-Se) = 2.537 Å, which are also in agreement with experimental results [72]. The close agreement between experimental and computational data supports our analysis.

TABLE 3.1 Computed DFT-based Descriptors of Compound $CuTiS_2$ and $CuTiSe_2$

Species	HOMO-LUMO Energy Gap (eV)	Hardness (eV)	Softness (eV)	Electro-negativity (eV)	Electro-philicity Index (eV)	Dipole Moment (Debye)
$CuTiS_2$	0.463	0.231	2.162	4.435	42.526	4.394
$CuTiSe_2$	0.471	0.235	2.124	5.038	53.918	3.548

TABLE 3.2 Quadrupole Moment of Compound $CuTiS_2$ and $CuTiSe_2$ in Debye-Ang

Species	Quad-xx	Quad-xy	Quad-xz	Quad-yy	Quad-yz	Quad-zz
$CuTiS_2$	−58.392	−3.616	0.0191	−54.302	−0.025	−57.969
$CuTiSe_2$	−62.640	0.095	0.000	−42.129	0.000	−73.621

TABLE 3.3 Comparison between Experimental and Computed Bond Lengths of Compound $CuTiS_2$ and $CuTiSe_2$ [71, 72]

Species	Theoretical Bond Length	Experimental Bond Length
d(Ti-S)	2.683	2.27
d(Cu-S)	2.256	2.35
d(Cu-Ti)	2.506	2.75
d(Cu-Se)	2.375	2.54
d(Se-Se)	2.538	3.63
d(Ti-Se)	2.537	2.54

3.4 CONCLUSION

The physical and chemical properties of semiconducting materials $CuTiS_2$ and $CuTiSe_2$ have been investigated by using DFT methodology. The DFT-based descriptors such as HOMO-LUMO energy gap, electronegativity; hardness, softness, electrophilicity index, and dipole moment have been computed. The result shows that compound $CuTiS_2$ is more reactive as compared to the compound $CuTiSe_2$. The compound $CuTiSe_2$ have high energy gap of 0.471 eV whereas, $CuTiS_2$ have energy gap of 0.463 eV. The compound having maximum value of HOMO-LUMO energy gap also have high value of molecular hardness, electronegativity, and electrophilicity index and least value of molecular softness. A close agreement between experimental and our computed bond length is observed from this study.

KEYWORDS

- **density functional theory**
- **highest occupied molecular orbital**
- **light emitting diodes**
- **local spin density approximation**
- **lowest unoccupied molecular orbital**
- **solar cells**

REFERENCES

1. Luque, A., & Marti, A., (1997). *Phys. Rev. Lett., 78*, 5014.
2. Shockley, W., & Queisser, H., (1961). *J. Appl. Phys., 32*, 510.
3. Luque, A., & Marti, A., (2001). *Prog. Photovoltaics, 9*, 73.
4. Wahnon, P., & Tablero, C., (2002). *Phys. Rev. B, 65*, 165115.
5. Palacios, P., Sanchez, K., Conesa, J. C., & Wahnon, P., (2006). *Phys. Stat. Sol., 203*, 1395.
6. Marti, A., Marron, D. F., & Luque, A., (2008). *J. Appl. Phys., 103*, 073706.
7. Wang, T., Li, X., Li, W., Huang, L., Ma, C., Cheng, Y., Cui, J., Luo, H., Zhong, G., & Yang, C., (2016). *Mater. Res. Express, 3*, 045905.
8. Green, M. A., (2001). *Prog. Photovoltaics: Res. Appl., 9*, 137.
9. Nozawa, T., & Arakawa, Y., (2011). *Appl. Phys. Lett., 98*, 171108.
10. Aguilera, I., Palacios, P., & Wahnon, P., (2008). *Thin Solid Films, 516*, 7055.
11. Aguilera, I., Palacios, P., & Wahnon, P., (2010). *Sol. Energy Mater. Sol. Cells, 94*, 1903.
12. Chen, P., Qin, M., Chen, H., Yang, C., Wang, Y., & Huang, F., (2013). *Phys. Stat. Sol., 210*, 1098.
13. Tablero, C., (2009). *J. Appl. Phys., 106*, 073718.
14. Shay, J. L., & Wernick, J. H., (1975). *Ternary Chalcopyrite Semiconductors: Growth, Electronic Properties, and Applications.* Pergamon, Oxford.
15. Ohmer, M. C., Randey, R., & Bairamov, B. H., (1998). *Mater. Res. Bull., 23*, 16.
16. Birkmire, R. W., & Eser, E., (1997). *Annu. Rev. Mater. Sci., 27*, 625.
17. Palacios, P., Sanchez, K., Conesa, J. C., Fernandez, J. J., & Wahnon, P., (2007). *Thin Solid Films, 515*, 6280.
18. Hashikuni, K., Suekuni, K., Usui, H., Chetty, R., Ohta, M., Kuroki, K., Takabatake, T., Watanabe, K., & Ohtaki, M., (2019). *Inorg. Chem., 58*, 1425.
19. Zhou, H. G., Chen, H., Chen, D., Li, Y., Ding, K. N., Huang, X., & Zhang, Y. F., (2011). *Acta Phys. Chim. Sin., 27*, 2805.
20. Ranjan, P., & Chakraborty, T., (2019). *Acta Chim. Slov., 66*, 173.
21. Ranjan, P., Dhail, S., Venigalla, S., Kumar, A., Ledwani, L., & Chakraborty, T., (2016). *Mat. Sci. Pol., 33*, 719.

22. Ranjan, P., Chakraborty, T., & Kumar, A., (2017). *Nano Hybrids, 17*, 62.

23. Ranjan, P., Chakraborty, T., & Kumar, A., (2017). *Physical Sciences Reviews, 2*, 2016012.

24. Ranjan, P., & Chakraborty, T., (2018). *Key Eng. Mater, 777*, 183.

25. Ranjan, P., Kumar, P., Chakraborty, T., Sharma, M., & Sharma, S., (2020). *Mater. Chem. Phys., 241*, 122346.

26. Ranjan, P., Kumar, A., & Chakraborty, T., (2016). *AIP Conf. Proc., 1724*, 020072.

27. Ranjan, P., Kumar, A., & Chakraborty, T., (2016). *Mat. Today Proc., 3*, 1563.

28. Ranjan, P., Kumar, A., & Chakraborty, T., (2016). *J. Phys.: Conf. Ser., 759*, 012045.

29. Ranjan, P., Kumar, A., & Chakraborty, T., (2016). *IOP Conf. Ser.: Mater. Sci. Eng., 149*, 012172.

30. Ranjan, P., Venigalla, S., Kumar, A., & Chakraborty, T., (2014). *New Front. Chem., 23*, 111.

31. Venigalla, S., Seema, D., & Ranjan, P., (2014). *New Front. Chem., 23*, 123.

32. Ranjan, P., & Chakraborty, T., (2013). *J. Int. Acad. Phys. Sci., 17*, 1.

33. Ranjan, P., Kumar, A., & Chakraborty, T., (2014). Computational Study of nanomaterials invoking DFT-based descriptors. *Environmental Sustainability: Concepts, Principles, Evidences, and Innovations* (pp. 239–242). ISBN: 978-9-383-08375-6. Excellent Publishing House, New Delhi, INDIA.

34. Ranjan, P., Venigalla, S., Kumar, A., & Chakraborty, T., (2015). A theoretical analysis of Bi-metallic Ag-Au$_N$ (N = 1–7) nanoalloy clusters invoking DFT-based descriptors. *Research Methodology in Chemical Sciences-Experimental and Theoretical Approaches.* CRC Press, Apple Academic Press, Taylor & Francis, USA, ISBN: 978-1-771-88127-2.

35. Dhail, S., Ranjan, P., & Chakraborty, T., (2017). Correlation of the experimental and theoretical study of some novel 2-phenazinamine derivatives in terms of DFT-based descriptors. *Crystallizing Ideas-The Role of Chemistry.* Springer, ISBN: 978-3-319-31758-8.

36. Ranjan, P., Chakraborty, T., & Kumar, A., (2017). Theoretical analysis: Electronic and optical properties of small Cu-Ag nano alloy clusters. *Computational Chemistry Methodology in Structural Biology and Material Sciences.* CRC Press, Apple Academic Press, Taylor & Francis, USA, ISBN: 978-1-771-88568-3.

37. Ranjan, P., Chakraborty, T., & Kumar, A., (2018). A theoretical study of bimetallic CuAu$_n$ (n = 1–7) nanoalloy clusters invoking conceptual DFT-based descriptors. *Applied Chemistry and Chemical Engineering* (Vol. 4). CRC Press, Apple Academic Press, Taylor & Francis, USA, ISBN: 9781771885874.

38. Ranjan, P., Chakraborty, T., & Kumar, A., (2018). Computational investigation of au doped Ag nanoalloy clusters: A DFT Study. *Innovations in Physical Chemistry: Monograph Series-Methodologies and Applications for Analytical and Physical Chemistry* (Vol. 3). CRC & Apple Academic Press, USA, ISBN-9781771886215.

39. Ranjan, P., Chakraborty, T., & Kumar, A., (2018). A density functional study of Ag-doped Cu nanoalloy clusters. *Innovations in Physical Chemistry: Monograph Series-Physical Chemistry for Engineering and Applied Sciences* (Vol. 4). CRC & Apple Academic Press, USA, ISBN: 9781771886277.

40. Ranjan, P., Chakraborty, T., & Kumar, A., (2018). Theoretical analysis: Electronic, Raman, vibrational, and magnetic properties of Cu$_n$Ag (n = 1–12) nanoalloy clusters.

Theoretical and Quantum Chemistry at the Dawn of 21ˢᵗ Century (pp. 1–34). CRC & Apple Academic Press, USA, ISBN: 9781771886826.

41. Shalini, P. R., & Chakraborty, T., (2019). Theoretical computation of periodic descriptors invoking periodic properties. *Chemical Sciences and Engineering Technology.* CRC & Apple Academic Press, USA, ISBN: 9781771887052.

42. Ranjan, P., Kumar, A., & Chakraborty, T., (2019). Structural, electronic, and optical properties of $AuCu_n^\lambda$ ($\lambda = 0, \pm1$; n = 1–8) nanoalloy clusters: A DFT study. *Research Notes on Nanoscience and Nanotechnology.* CRC & Apple Academic Press, USA.

43. Tandon, H., Shalini, P. R., Suhag, V., & Chakraborty, T., (2019). A Review on the computational study of carbon nanotubes. *Carbon Nanotubes and Nanoparticles-Current and Potential Applications.* CRC & Apple Academic Press, USA, ISBN: 9781771887342.

44. Ranjan, P., Chakraborty, T., & Kumar, A., (2019). The study of physico-chemical properties of bimetallic CuAun (n = 1–8) nanoalloy clusters. *Research Notes on Nanoscience and Nanotechnology.* CRC & Apple Academic Press, USA.

45. Ranjan, P., Chakraborty, T., & Kumar, A., (2019). Conformational study of bimetallic trimers Cu-Ag nanoalloy clusters. *Research Notes on Nanoscience and Nanotechnology.* CRC & Apple Academic Press, USA.

46. Ranjan, P., Chakraborty, T., & Kumar, A., (2019). Conformational analysis of Ag-Au nanoalloy clusters: A CDFT approach. *Composite Materials Engineering Modeling and Technology.* CRC & Apple Academic Press, USA.

47. Ranjan, P., & Chakraborty, T., (2020). Theoretical analysis of Au-V nanoalloy clusters: A density functional approach. *Materials Physics and Chemistry: Applied Mathematics and Chemo-Mechanical Analysis.* CRC Press, Apple Academic Press, Taylor & Francis, USA, ISBN: 9781771888677 (Accepted-In Press,).

48. Ranjan, P., & Chakraborty, T., (2020). A DFT study of Cu_nFe (n = 1–5) nanoalloy clusters. *Materials Physics and Chemistry: Applied Mathematics and Chemo-Mechanical Analysis.* CRC Press, Apple Academic Press, Taylor & Francis, USA, ISBN: 9781771888677 (Accepted-In Press,).

49. Parr, R. G., & Yang, W., (1989). *Density Functional Theory of Atoms and Molecules.* Oxford, New York.

50. Dobson, J. F., Vignale, G., & Das, M. P., (1998). *Electron Density-Functional Theory.* Recent Progress and New Directions, Plenum, New York.

51. Hafner, J., Wolverton, C., & Ceder, G., (2006). *MRS Bull., 31,* 659.

52. Gaussian, 03, Revision, C.02, Frisch, M. J., Trucks, G. W., Schlegel, H. B., Scuseria, G. E., Robb, M. A., et al., (2004). Gaussian, Inc.: Wallingford, CT.

53. Ghosh, D. C., & Bhattacharyya, S., (2004). *Int. J. Mol. Sci., 5,* 239.

54. Fujimoto, H., Kato, S., Yamabe, S., & Fukui, K., (1974). *J. Chem. Phys., 60,* 572.

55. Fujimoto, H., Kato, S., Yamabe, S., & Fukui, K., (1974). *J. Am. Chem. Soc., 96,* 2024.

56. Sanderson, R. T., (1951). *Science, 114,* 670.

57. Sanderson, R. T., (1952). *Science, 116,* 41.

58. Sanderson, R. T., (1955). *Science, 121,* 207.

59. Sanderson, R. T., (1952). *Science, J. Chem. Edu., 29,* 539.

60. Sanderson, R. T., (1954). *Science, Science, 31,* 238.

61. Sanderson, R. T., (1960). *Chemical Periodicity.* Reinhold Publishing Corporation: New York.

62. Ghosh, D. C., (1984). *Indian J. Pure Appl. Phys., 22,* 346.
63. Ghosh, D. C., (1989). *Indian J. Pure Appl. Phys., 27,* 160.
64. Parr, R. G., Donnelly, R. A., Levy, M., & Palke, W. E., (1978). *J. Chem. Phys., 68,* 3801.
65. Parr, R. G., & Pearson, R. G., (1983). *J. Am. Chem. Soc., 105,* 7512.
66. Cedilo, A., Chattaraj, P. K., & Parr, R. G., (2000). *Int. J. Quan. Chem., 77,* 403.
67. Chattaraj, P. K., Nath, S., & Sannigrahi, A. B., (1994). *J. Phys. Chem., 98,* 9143.
68. Sannigrahi, A. B., & Nandi, P. K., (1994). *J. Mol. Struc. (Theo Chem.), 307,* 99.
69. Xiao, H., Kheli, J. T., & Goddard, III. W. A., (2011). *J. Phys. Chem. Lett., 2,* 212.
70. Klepp, K. O., & Gurtner, D., (1996). *J. Alloys Compd., 243,* 19.
71. Jurelo, A. R., Ribeiro, R. A. P., De Lazaro, S. R., & Monteiro, J. F. H. L., (2018). *Phys. Chem. Chem. Phys., 20,* 27011.

CHAPTER 4

Synergistic Effect of *E. crassipes* Biomass/Chitosan for As (III) Remediation From Water

PANKAJ GOGOI,[1] PAKIZA BEGUM,[2] KAUSTUBH RAKSHIT,[3] and TARUN K. MAJI[4]

[1]*Department of Chemistry, Sipajhar College, Darrang – 784145, Assam, India, E-mail: pankaj5@yahoo.com*

[2]*Department of Chemistry, Indian Institute of Technology Guwahati, North Guwahati, Kamrup – 781039, Assam, India, E-mail: pakiza@iitg.ac.in*

[3]*Center for the Environment, Indian Institute of Technology Guwahati, North Guwahati, Kamrup – 781039, Assam, India, E-mail: kaustubhrakshit@gmail.com*

[4]*Department of Chemical Sciences, Tezpur University, Napaam, Tezpur, Sonitpur – 784028, Assam, India, Tel.: +91 3712 267007, Ext: 5053, Fax: +91 3712 267005, E-mail: tkm@tezu.ernet.in*

ABSTRACT

Current study illustrates the synergistic influence of amalgamating chitosan with dried *E. crassipes* root powder for As (III) remediation from water; later being inexpensive and abundant biomaterial. The composite was found to be very effective in removing As (III) to <10 µg/L, the permissible limit prescribed by the World Health Organization (WHO). Interactions among component materials in the composite and with adsorbed arsenic were analyzed and confirmed by different physicochemical/spectroscopic tools and further verified by density functional theory (DFT) calculations.

Physical parameters such as material dose, treatment time, and initial arsenic concentration, could alter efficiency of the material. For a test sample containing 0.4 mg/L of arsenic, 3 g/L of the material could effectively bring down As (III) concentration below acceptable limit. Langmuir adsorption isotherm could reasonably explain the sorption pattern and maximum adsorption capacity was found to be 7.11 mg of arsenic/g. The sorption was chemical in nature and was governed by a pseudo-second-order kinetic model.

4.1   INTRODUCTION

Prevalence of arsenic in drinking water is threatening millions of people's health over the world. It leads to many skin disorders, kidney troubles, heart diseases, diabetes, paralysis, etc., and most interestingly it is in focus as a cancer-causing agent in recent times [1–4]. People especially from the developing countries are mostly affected by its contamination as they are unable to avail efficient cost-effective tools and techniques to get relief from this deadly contaminant [5]. Besides certain acute health hazards, prolonged intake of arsenic-contamination in water causes many serious health effects, called arsenicosis [6, 7]. Cost-effective, simple methods using easily, and widely available materials therefore, can bring some sigh of relief to those people [8]. Adsorption is one of such widely used concept when arsenic is present in trace level [9].

Chitosan (CT) biopolymer has been identified as an efficient adsorber for various arsenic species [10]. Many CT based materials have been successfully employed for the removal of Cr (VI), Cu (II), Cd (II), As (III), As (V), etc. [11, 12]. CT is bio-compatible, non-toxic, and bio-degradable polymer as compared to synthetic polymers which can be attributed to the chemically reactive functional groups like hydroxyl, acetamido, amino groups, etc., which makes CT an edge over others [4]. But CT is expensive and this limits its use in water treatment. Nevertheless, its excellent sorption ability encourages researchers to make it useful by adopting certain measurements to reduce cost [13]. In this regard different biomasses which mainly comprise of cellulosic material having more or less similar structure to CT may bring some significant outcomes. Combination of those with CT has many aspects to reduce cost without compromising sorption efficiency [14].

Water hyacinth (*E. crassipes*), in its living form or dried biomass of its different parts have been established as good sorbent for water treatment. It is made up of mainly cellulosic and lignocellulosic components. Dried biomass of different parts of this aquatic weed in their unmodified/modified form has been explored as good sorbent for many toxic metal ions including arsenic [15, 16].

Although dried biomass of water hyacinth roots powder is cost-effective for water treatment but in order to make it useful it is necessary to remove the turbidity and flavor that persists after treatment. Jiufang et al. have prepared a cellulose-CT gel complex via intermolecular inclusion interaction which has self-healing ability [8]. The turbidity and leaching can be minimized by coating the powder with polymeric material. This may not only remove the turbidity but also prevent the agglomeration and thus may enhance its efficacy. Keeping these in view, the present study has been undertaken. The present study aims to prepare a composite of root powder with CT biopolymer stabilized by glutaraldehyde cross-linker and utilized the advantages of both the materials to develop a combined system having superior property which will be cheap in cost. Effect of different parameters on efficacy of arsenic removal have been studied and optimized. A model describing the different interactions during the preparation of CT-water hyacinth root powder (WHRP)-glutaraldehyde composite and its binding with arsenic is presented with the help of density functional theory (DFT) and verified experimentally with FTIR.

4.2 EXPERIMENTAL

4.2.1 MATERIALS

Roots of water hyacinth were collected from local pond, Tezpur University campus. Sodium meta-arsenite ($NaAsO_2$) and CT (M_v~5.96×10^5, DDA~75) were purchased from Sigma. NaOH was purchased from Merck, Mumbai, India. De-ionized water was obtained from Milli-Q water purification system (Millipores.A.S.67 120 MOLSHEIM, FRANCE). The rest reagents were of analytical grade and used without further purification.

4.2.1.1   PREPARATION OF CTRP

The collected roots of water hyacinth were initially washed several times with water. Then it was kept in 0.1 M NaOH solution for 24 hours and again washed several times with surfactant. Washing with NaOH solution helped reducing the dirt, other soluble parts and lignin from the root powder. Finally, it was washed with distilled water and allowed to dry at about 40°C in hot air oven. The dried material (WHRP) thus obtained was grinded to powder and sieved to 100 mesh size and kept for subsequent uses. Simultaneously a CT solution was prepared by dissolving a requisite amount of it in 1% (v/v) acetic acid solution. Then different variations of WHRP (in 1:0.25, 1:0.5 and 1:1 ratio w.r.t. CT content) were added to it to prepare three composite systems. The mixture was then stirred vigorously for half an hour. NaOH solution was then added to the above mixture till the completion of CT precipitation followed by the addition of glutaraldehyde solution under cold condition. The temperature was then gradually increased to 50–55°C and allowed to crosslink for another 3 h. Finally, it was filtered and washed with de-ionized water until it became neutral and then the sample was dried in a hot air oven at about 60°C.

4.2.2   MEASUREMENTS

4.2.2.1   FT-IR ANALYSIS

Fourier transformed infrared spectrophotometer (FTIR) (NICOLET Impact I-410) was used to study the presence of different functional moiety in different component materials and their probable interactions in composite formation and binding process. The study was carried out in the range of 500–4000 cm^{-1}.

4.2.2.2   THERMO GRAVIMETRIC ANALYSIS (TGA)

TGA was recorded for all the samples using a thermogravimetric analyzer (Metler TA 400) at a heating rate of 10°C/min up to 600°C under N_2 atmosphere.

4.2.2.3 SEM-EDX ANALYSIS

Scanning electron microscope (JEOL 6390 LV) was used to study the surface morphologies and EDX was carried out to study the elemental composition of the samples before and after As (III) treatment.

4.2.2.4 pH MEASUREMENT

The pH of the arsenic solutions was adjusted using either NaOH (0.1 M) and/or HCl (0.1 M) solutions as and when required and measured by cyber scan pH 510 (Eutech) instrument.

4.2.2.5 ATOMIC ABSORPTION SPECTROSCOPIC ANALYSIS

Analyst 200 Atomic Absorption Spectrophotometer (AAS, Perkin Elmer) was used to measure the As (III) concentration. All the measurements were based on integrated absorbance and performed at 193.7 nm by using a quartz tube analyzer (Perkin Elmer) at an atomization temperature of 2000 K.

4.2.3 BATCH ADSORPTION TEST

A stock solution containing 1000 mg/L of As (III) was prepared by dissolving a requisite amount of sodium meta-arsenite in 1000 mL de-ionized water. The stock solution was diluted in steps as required to obtain standard solutions containing 0.1–0.8 mg/L of As (III) solution. Batch experiment was carried out in a mechanical shaker at an agitation speed of 160 rpm at neutral pH in 100 mL conical flasks, at room temperature. After agitation, the suspension was centrifuged, filtered, and the arsenic concentration left in the aliquot was determined by atomic absorption spectrophotometer. Initially, for a targeted solution of 0.4 mg/L, the effect of doses of the different materials under investigation was carried out. Then the influence of agitation time on percent removal of As (III) was tested keeping the concentration of initial ion and the agitation time unchanged for that optimized dose. Effect of initial ion concentration on As (III) removal rate was carried out in the concentration range of 0.1–0.7 mg/L for the optimized dose and time of treatment.

The equilibrium adsorption capacity was calculated using the Eqn. (1) below:

$$q_e = \frac{(C_o - C_e)V}{M} \tag{1}$$

where, q_e (mg/g) is the equilibrium adsorption capacity, C_o and C_e are the initial and equilibrium concentration (mg/l) of As (III) in solution, V(L) is the volume and M(g) is the weight of the adsorbent.

Adsorption kinetic studies and isotherm studies were carried out with four different initial concentrations as shown in Tables 4.3 and 4.4, respectively, while maintaining the adsorbent dosage at 3 g/L. Langmuir and Freundlich models were applied to study the adsorption isotherm and different constants were generated. Pseudo first order and second order were applied to study the mechanism of sorption.

The Linear form of a simple pseudo first order kinetic model is represented as [17]:

$$\log_{10}(q_e - q_t) = \log_{10} q_e - \frac{k_{ad}}{2.303t} \tag{2}$$

where, q_t is the amount of arsenic adsorbed (mg/g) at time t, q_e is the amount of As (V) adsorbed (mg/g) at equilibrium and K_{ad} is the rate constant of adsorption (min^{-1}). The values of k_{ad} and q_e were calculated from the slope and intercept of the respective linear plots. The correlation coefficients (R^2) were computed and the values are shown in Table 4.3.

The linear form of type I pseudo-second-order kinetic models is represented by the following equation [18].

$$\frac{t}{q_t} = \frac{1}{h} + \frac{t}{q_e} \tag{3}$$

where, $q_t = q_e^2 kt/(1+q_t kt)$, the amount (mg/g) of As (V) adsorbed on the surface of material at any time t, k being pseudo-second-order rate constant (gmg^{-1}min^{-1}), q_e is the amount adsorbed at equilibrium, and $h = kq_e^2$ (mgg^{-1} min^{-1}) is the initial sorption rate.

Langmuir and Freundlich isotherm models were applied for studying the sorption pattern. The Langmuir isotherm is represented by the following equation [12]:

$$\frac{1}{q_e} = \frac{1}{q_m b} \frac{1}{C_e} + \frac{1}{q_m} \tag{4}$$

where, q_e is the equilibrium quantity adsorbed (mg/g), C_e is the equilibrium concentration (mg/L), q_m is the maximum adsorption capacity (mg/g) and 'b' is the Langmuir constant.

The separation factor, R_L an important parameter indicating the favorability of the adsorption based on the Langmuir equation is calculated using equation (5).

$$R_L = \frac{1}{(1 + bC_0)} \tag{5}$$

where, b, and C_0 have their common meaning. The value of R_L indicates the type of adsorption either to be unfavorable ($R_L > 1$), linear ($R_L = 1$), favorable ($0 < R_L < 1$) or irreversible ($R_L = 0$).

The linear form of Freundlich adsorption equation is represented by the equation [19]:

$$\ln q_e = \ln k_f + \frac{1}{n} \ln C_e \tag{6}$$

where, q_e is the adsorbed amount (mg/g), C_e is the equilibrium arsenic concentration (mg/L), k_f (mg/g) is the Freundlich constant related to adsorption capacity and 'n' is a constant related to energy of intensity of adsorption.

4.2.4 DENSITY FUNCTIONAL THEORY (DFT) STUDIES

All density functional calculations were carried out using DMol^3 package [20, 21] utilizing the nonlocal exchange-correlation functional of Perdew et al., (PW91) exchange-correlation functional at the generalized gradient approximation (GGA) level [22]. The van der Waals (vdW) interactions were taken into consideration while performing theoretical calculations, as polymeric structures bind through vdW interactions. Dispersive forces, or vdW forces, result from the interaction between fluctuating multipoles without requiring the overlap of electron densities [23]. Therefore, dispersion corrected density functional theory (DFT-D) for geometry optimization and frequency analysis was utilized. DFT-D approaches to treat vdW interactions was employed using the Ortmann, Bechstedt, and Schmidt (OBS) correction to VWN and double numerical plus polarization (DNP) basis set was chosen for the calculations [17, 18, 24]. The geometries of reactants and the cross-linked product were optimized without imposing any symmetry constraints using all electron spin-unrestricted calculation, indicating electronically open shell system. For the heavy metal Arsenic, the valence electrons were described by double numerical basis set with polarization function and the core electrons were described with local pseudo-potential (VPSR) which accounts for the scalar relativistic effect expected to be significant for heavy metal elements [25]. The structures were fully relaxed and positive vibrational frequencies confirmed the complexes to be at energy minima.

4.3   RESULTS AND DISCUSSION

4.3.1   CHARACTERIZATION

4.3.1.1   FT-IR ANALYSIS

Figure 4.1. [A] Represents the FT-IR spectra of CT (i), WHRP (ii), CTRB (iii), CTRP (iv), and CTRP-As (v) respectively in the first part.

The characteristics absorption bands shown by all the spectra near 3435 cm^{-1} and 2929 cm^{-1} were due to O-H, N-H stretching, and due to CH_2 stretching vibrations [26, 27]. In WHRP (curve-b), band near 1642 cm^{-1} may be assigned to the C=O stretching vibration of carboxylate in lignocellulosic part. Band at 1411 cm^{-1} was due to the scissoring modes of $-CH_2$ stretching in methylene chains and at 1031 cm^{-1} was assigned for C-O stretching, typical of cellulose [14, 28]. Absence of band in the range 1500–1600 cm^{-1}, characteristics of aromatic C=C in WHRP indicated that the lignin part was removed during the NaOH treatment process [29]. The intense band near 1642 cm^{-1} might be assigned to deformation mode of NH_2 in CT or due to the bending vibration resulting from H-O-H intermolecular linkages. Similarly, in the physical blend of CT and WHRP (Figure 4.1 [A] (c)), increase in intensity near 1642 cm^{-1} might be due to the increase in the amount of –OH functionality. Another important band around 1164 cm^{-1} due to C-N stretching was observed in the spectrum of CT and all other materials containing CT. Change in intensity of this peak (Figure 4.1 [A] (e)) along with that of 1642 cm^{-1} (may be due to the formation of N-C=O bond during cross-linking) confirmed the formation of new chemical bond of glutaraldehyde with the different polymeric moieties. No characteristic band appeared in the spectrum for aldehydic group utilized in the cross-linking process.

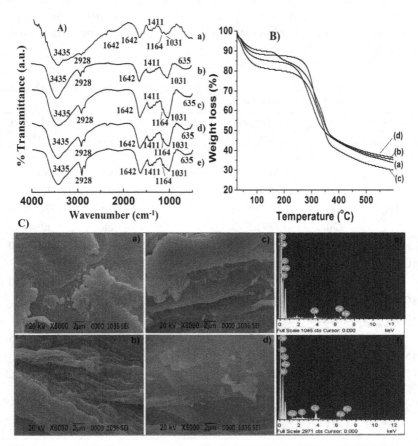

FIGURE 4.1.   [A] FT-IR spectra of (a) CT, (b) WHRP, (c) CTRB, (d) CT:RP/1:2, (e) CT:RP/1:2-As; [B] TGA thermo grams of (a) CT, (b) WHRP, (c) CTRB, (d) CT:RP/1:2; [C] SEM images of (a) CT, (b) WHRP, (c) CTRP (d) CTRPAs and EDX spectra of (e) CTRP, (f) CTRP-As.

The intensity of the peaks appeared at 1031 cm^{-1} and around 636 cm^{-1} increased after arsenic sorption on to the material but that of 1642 cm^{-1} decreased during the process (Figure 4.1 [A] (e)). Those intensity measurements were carried out with respect to the –CH$_2$ stretching frequency. This suggested the possible binding of arsenic onto the surface of cross-linked root powder. The decrease in intensity near 1642 cm^{-1} was due to the utilization of NH$_2$ and carboxylate functionalities in the binding process. In the contrary, the increase in intensities in the former cases was due to the contribution of newly formed O-As bond during the sorption. This indicated successful sorption of arsenic on the composite material (Table 4.1).

TABLE 4.1 Experimental and Theoretical Infrared Spectral Data for the Optimized Complexes, v is the Vibrational Frequency in cm^{-1}

Assignment		Frequency (cm^{-1})	Assignment		Frequency (cm^{-1})
Complex 1(d)			Complex 1(e)		
v(C-O)$_{stretching}$	Exp.[a]	1060–1175	v(C-O)$_{stretching}$	Exp.	1060–1175
	Calc.	1078.8		Calc.	1062.1
v(O-H)$_{bending}$	Exp.[b]	1375–1400	v(O-H)$_{bending}$	Exp.	1375–1400
	Calc.	1391.5		Calc.	1394.2
Complex 1(f)			v(N-H)$_{stretching}$	Exp.	3400–3500
v(C-O)$_{stretching}$	Exp.	1060–1175		Calc.	3430.1
	Calc.	1067.7	v(N-H)$_{bending}$	Exp.	1400–1600
v(O-H)$_{bending}$	Exp.	1375–1400		Calc.	1626.3
v(N-H)$_{stretching}$	Calc.	1393.8	*Complex 1(g)*		
	Exp.[c]	3400–3500	v(C-O)$_{stretching}$	Exp.	1060–1175
v(N-H)$_{bending}$	Calc.	3447.7		Calc.	1071.1
	Exp.[d]	1400–1600	v(O-H)$_{bending}$	Exp.	1375–1400
	Calc.	1632.8		Calc.	1396.8
v(As-O)$_{stretching}$	Exp.[e]	530–700	v(As-O)$_{stretching}$	Exp.	530–700
	Calc.	627.5		Calc.	726.2

DFT calculations were carried out in order to further confirm the structure and study the electronic properties of the novel As/CT-cellulose complex. The DFT optimized geometry of As (III), cross-linked polymeric materials and their binding with the former are shown in Figure 4.2. The vibrational frequencies obtained from the optimized geometries are shown in Table 4.1 which was found to be in accordance with the experimental results. The hydrogen bond energies for the adsorption of the As (III)-hydroxide on the CT-cellulose complex was found to be 39.16 kcal.mol^{-1} suggesting the bond to be strong and mostly covalent in nature [30]. Whereas, for the adsorption of As$_4$O$_6$ on cellulose-cellulose complex, energy of the hydrogen bond was found to be 8.34 kcal.mol^{-1}, indicating that the bond formed is moderate in strength and electrostatic in nature.

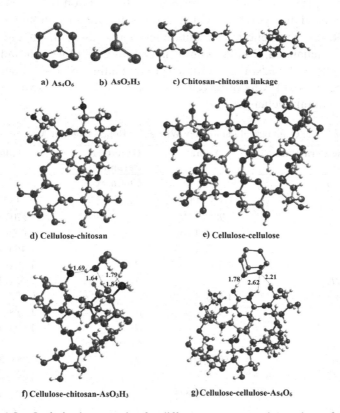

a) As$_4$O$_6$ b) AsO$_3$H$_3$ c) Chitosan-chitosan linkage

d) Cellulose-chitosan   e) Cellulose-cellulose

f) Cellulose-chitosan-AsO$_3$H$_3$ g) Cellulose-cellulose-As$_4$O$_6$

FIGURE 4.2 Optimized geometries for different reactants and complexes formed with the polymers.

The geometric parameters of final optimized structures are tabulated in Table 4.2. The structures with their bond lengths marked are shown in the supplementary information in Figure 4.2(a–g). From Table 4.2 it was observed that the complex *(f)* which was formed by the sorption of *(b)* on *(d)*, the O-H^a bond near to the NH$_2$ group shortened after arsenic sorption, whereas, the length for other OH^b bond remain unchanged. The

N-H bond also shortened whereas, lengthening of C-N bond took place compared to the untreated polymeric system. The As-O as well as AsO-H bonds got elongated in complex *(f)*. This confirmed that the sorption of As (III) caused some changes in the geometry of the complex particularly in those atoms where direct interactions occurred.

For complex *(g)* which was formed by the sorption of *(a)* on *(e)*, the O-H[a] bonds of the polymer which formed hydrogen bonding with the As_4O_6 cluster was lengthened but other bonds away from the cluster remained unchanged. Similar lengths in case of O-C-O were found for the pure as well as the adsorbed complex. Lengthening of As-O[e] bond of the lower ring of As_4O_6 was observed, the other As-O[f] bonds away from the polymer remain unchanged.

TABLE 4.2 The Geometrical Parameters Where *d* is Bond Length in Å

Geometric Parameters	Values	Geometric Parameters	Values
Complex (d)		**Complex (f)**	
d_{O-H}	1.035–1.040[a]	d_{O-H}	1.008[a]
	0.986–0.995[b]		0.970–0.985[b]
d_{N-H}	1.024–1.048	d_{N-H}	1.022–1.023
d_{C-N}	1.452–1.454	d_{C-N}	1.460–1.468
d_{C-O}	1.400–1.448	d_{C-O}	1.422–1.437
Complex (b)		d_{As-O}	1.831–1.881
d_{As-O}	1.838–1.858	d_{AsO-H}	0.993–1.020
d_{AsO-H}	0.975–0.978	$d_{AsOH \cdots NH}$	1.688–1.794
		$d_{OH \cdots OHAs}$	1.693–1.837
Complex (e)		**Complex (g)**	
d_{O-H}	0.970–0.985	d_{O-H}	0.981–0.992[c]
			0.975–0.986[d]
d_{C-O}	1.396–1.445	d_{C-O}	1.387–1.452
Complex (a)		d_{As-O}	1.856–1.859[e]
d_{As-O}	1.850–1.851		1.842–1.850[f]
		$d_{OH \cdots OAs}$	2.210–2.617
		$d_{As \cdots OH}$	2.860

[a] OH bond length of those atoms which are near to the NH$_2$ group.
[b] OH bond length of all other bonds away from NH$_2$.
[c] OH bond length which form links to the As_4O_6 cluster.
[d] OH bonds away from the As_4O_6 cluster.
[e] As-O bonds of the lower ring of As_4O_6 cluster which is adsorbed on the polymeric system.
[f] As-O bonds away from the polymeric system.

4.3.1.2   TGA ANALYSIS

Thermo grams of CT, WHRP, chitosan root powder physical blend (CARB), and chitosan root powder composite (CTRP) were shown in Figure 4.1 [B]. In all the thermograms, the initial weight loss at around 100°C was due to evaporation of moisture. The major weight loss occurred in the range 230–360°C. The weight loss was least for the composite followed by CT and WHRP. It was highest for the blend. Also, the percentage of char was more in CTRP compared to CARB, CT or WHRP. For the physical blend, the degradation temperature was shifted to low temperature as compared to the individual materials. Cellulose and CT contain large amount of intra-and intermolecular hydrogen bonds in their structure [31, 32]. These intra and intermolecular hydrogen bonds may be weakened during the physical blending and as a result degradation took place at lower temperature. Similar results were observed by Cai and Kim when experimented with different variations of CT w.r.t. cellulose [33]. In contrast to this, the two materials when crosslinked with glutaraldehyde showed significant improvement in the thermal stability due to the functioning of extensive chemical bonding among the different polymeric chains.

4.3.1.3   SEM-EDX ANALYSIS

Figure 4.1 [C] (a-d) represents the SEM images of CT, WHRP, CTRP, and arsenic adsorbed CTRP composite respectively. In the micrograph for CT (a), a non-porous, smooth surface was observed having numbers of crystallites which resembles to those of results reported by Kumar and Koh [34]. In contrast to this, the SEM image of the WHRP showed numbers of fibriller crystallites. These structures were mainly due to the cellulosic component of the WHRP. But in the combined system, further changes in the surface morphology were observed as compared to the WHRP. Although some irregularity in the structure was observed after incorporation of WHRP in CT, the new structure showed homogenous distribution of the components which was further supported by the degradation pattern in TGA study. This indicated the successful interaction among the components in the composite.

The surface of arsenic adsorbed CTRP composite (d) appeared smooth which might be due to the sorption of the arsenic on the material.

EDX spectrum of CTRP (Figure 4.1 [C] (e)) showed the presence carbon, nitrogen, and oxygen which are the major elements of glucose and glucose amine of CT and cellulose of the components of the composite material. Along with those, metals like calcium, iron, etc., were also observed which may be attributed to WHRP unit [35]. The presence of arsenic along with other metals in the EDX spectrum of CTRP-As (Figure 4.1 [C] (f)) indicated the successful adsorption of arsenic on the surface of the composite system.

4.3.2   ADSORPTION RESULTS

4.3.2.1   EFFECT OF MATERIAL DOSE AND COMPOSITION ON REMOVAL OF ARSENIC

To study the effect of dose and composition on removal rate of arsenic, the arsenic solution (0.4 mg/L) was treated with different doses of material and stirred at about 160 rpm for 5 h. The test was carried out for different materials having different compositions of CT and WHRP viz. CT:RP/1:0, CT:RP/0:1, CT:RP/1:4, CT:RP/1:2, CT:RP/1:1 respectively and material dose was varied from 1–6 g/L. The removal rate increased with the increase in material dose and the sorption capacities of the different materials were found to be different as shown in (Figure 4.1 [C] (a)). The sorption efficiency was highest for glutaraldehyde crosslinked CT and least for the WHRP. The overall removal rate of CT:RP/1:2 and CT:RP/1:1 were comparable to that of CT alone. The turbidity of the solution that persists after treatment with CT/RP was negligible compared to that of WHRP alone (25 NTU). Also, the composite materials showed similar removal efficiency as that of CT, although the amount of CT in the former is significantly less. Similarly, to get similar efficacy, the amount of composites required was very less compared to that of WHRP alone (Figure 4.3).

The enhanced efficacy of the composite system was due to the synergistic effect of both the polymeric materials, i.e., CT, and WHRP present in the composite. During the mixing of the two polymeric materials a number of inter and intramolecular H-bonding sites might get broken resulting into formation of a number of free active sites for sorption. The amount of material required for maximum removal of arsenic (arsenic solution under investigation) was found to be 3 g/L. By considering efficiency, turbidity due to the material and cost, the CT:RP/1:2 composite was selected for subsequent investigations.

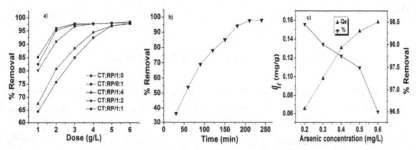

FIGURE 4.3 Effect of (a) material dose on removal rate, (b) treatment time, and (c) initial ion concentration on sorption efficiency.

A dose of 3 g/L of the chosen material found to be sufficient for bringing down the solution concentration below 10 µg/L. Beyond that dose, the removal efficiency was not so much significant. Therefore, a material dose of 3 g/L was taken for further study. With the increase in material dose, the number of active functional/or binding/and surface sites increased and hence the sorption increased. Influence of further increase in material dose was not so significant. Because of the presence of very small amount of As (III) ions in solution which might instantly attain equilibrium with the adsorbed As (III) ions present on the surface of the material.

4.3.2.2 EFFECT OF TREATMENT TIME ON REMOVAL EFFICIENCY

Figure 4.3(b) shows the effect of treatment time on removal efficiency of As (III) (in the range of 30–240 min). 100 mL of As (III) solutions (0.4 mg/L) were taken in eight conical flasks each containing 3 g of CT:RP/1:2 per liter of solution. These mixtures were allowed to agitate at a speed of 160 rpm and each of the conical flasks was taken out from the shaker at a regular interval of half an hour. The supernatants were centrifuged, filtered, and aliquots were analyzed for As (III).

With the increase in agitation time, the equilibrium adsorption capacity of the material increased up to a certain time (210 min) beyond that it remained almost constant. Increase in agitation time allowed more functional sites to bind with As (III) ions. But as sorption being a reversible process, after certain time, equilibrium was established between the adsorbed arsenic on the material and the arsenic present in the solution.

4.3.2.3   EFFECT OF INITIAL ION CONCENTRATION

The effect of initial ion concentration on adsorption capacity/removal rate of arsenic was studied in the concentration range of 0.2–0.5 mg/L for a material dose of 3 g/L for 210 min. Figure 4.3(c) shows the variation of adsorption capacity/removal rate of arsenic with change in initial arsenic concentration. It was observed that the equilibrium sorption capacity (mg/g) was increased with the increase in initial arsenic concentrations while percent removal decreased. The increase in sorption capacity was due to the increase in concentration gradient of arsenic at the interface of adsorbent and the arsenic solution. Again, due to the increase in arsenic concentration, the equilibrium between the arsenic ion present in solution and the adsorbent surface was delayed. As a consequence, the removal rate decreased.

4.3.2.4   ADSORPTION KINETICS

Figure 4.4 [A] represents the variation of q_t (mg/g) of arsenic overtime for four different initial arsenic ion concentrations viz. 0.2, 0.3, 0.4 and 0.5 mg/L. The q_t values were found to increase with the increase in the initial arsenic concentrations. This might be due to high concentration gradient of arsenic at the interface of sorbent and solution. The plots of $\log(q_e - q_t)$ vs. t and t/ q_t vs. t as per Eqs. (2) and (3) gave straight lines from which the different kinetic parameters for pseudo-first-order and pseudo-second-order were evaluated and given in Table 4.3.

The correlation coefficients (R^2) were computed for the all models and the values are shown in Table 4.3. The curve for pseudo-second-order kinetics exhibited higher correlation co-efficient compared to the other models indicating the preference for a monolayer chemisorptions pattern.

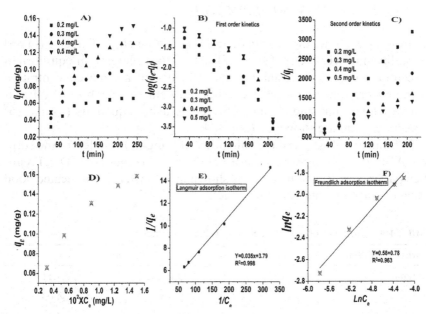

FIGURE 4.4 *(a-c) Q_t vs t curve and kinetic models, (d-f) isotherm models for sorption of As(III) on CT:RP/1:2.*

TABLE 4.3 Parameters for Different Kinetic Models for Sorption of Arsenic on CT:RP/1:2

C_o (mg/L)	First Order Kinetic Model				Second-Order Kinetic Model			
	Q_e (exp) mg/g	$K_1 \times 10^{-2}$ (min^{-1})	Q_e (mg/g)	R^2	$K_2 \times 10^{-2}$ g mg min^{-1}	Q_e (Cal) (mg/g)	$h \times 10^{-3}$	R^2
0.2	0.066	2.41	0.087	0.926	29.7	0.079	1.87	0.995
0.3	0.098	2.47	0.156	0.920	14.2	0.126	2.25	0.991
0.4	0.131	2.53	0.319	0.808	5.86	0.186	2.04	0.995
0.5	0.143	2.49	0.371	0.769	4.30	0.221	2.10	0.998

4.3.2.5   ADSORPTION ISOTHERM STUDY

Figure 4.4 [D] shows the growth of adsorbed amount (q_e) of arsenic (mg/g) at equilibrium over the surface CA/10RP against increase in equilibrium concentrations (C_e). To establish the influence of arsenic concentration on the adsorption process, the equilibrium data was analyzed by linear forms of Langmuir and Freundlich models using equations (4) and (6) respectively and are shown in Table 4.4.

R_L value (0.64) showed the suitability of Langmuir model for the sorption. Also, q_e vs C_e showed linearity of the graph (Figure 4.4 [D]). Freundlich parameters R^2 and n values as shown in Table 4.4 also indicated good fitting of sorption data with the model.

TABLE 4.4   Various Isotherm Models and Related Parameters

Isotherm Model	Parameters	
Langmuir	Q (mg/g)	7.11
	b	4.03
	R_L	0.64
	R^2	0.998
	k_f (mg/g)	1.69
Freundlich	n	1.80
	R^2	0.963

R^2 values for both Langmuir and Freundlich models showed good co-relations of the data and the sorption could be reasonably explained by the models under consideration. R^2 values for Langmuir model was higher than that of Freundlich model. Thus the preferential order for sorption for the applied models is as follows-Langmuir > Freundlich.

4.4   CONCLUSION

Combined system of WHRP and CT stabilized with glutaraldehyde could be successfully utilized for arsenic remediation process. The preparation of the material was very simple and very good interactions occurred among the different components. The sorption of As (III) over those materials was enormously explained by different analytical tools. Moreover, that information were further supported by the theoretical study.

Variation of composition of component materials significantly influenced the sorption process. Different parameters like material dose, treatment time, initial arsenic concentration, etc., also had remarkable influence on the efficacy of the prepared material on arsenic sorption. The sorption capacity of the material was found to be a function of initial arsenic concentration and increased with the increase in ion concentration. For an initial As (III) concentration of 0.4 mg/L a material dose of 3 g/L was found to be sufficient to bring the concentration below acceptable limit.

The sorption preferred a monolayer chemisorption pattern as described by Langmuir adsorption isotherm. The sorption of As (III) over the material was mainly via the different functional moiety. The findings were explained by pseudo-second-order kinetic model and confirmed by DFT study.

KEYWORDS

- **chitosan**
- **density functional theory**
- **double numerical plus polarization**
- **isotherm**
- **synergistic effect**
- **water hyacinth**

REFERENCES

1. Mandal, B. K., & Suzuki, K. T., (2002). Arsenic round the world: A review. *Talanta, 58*, 201–235.
2. Duker, A. A., Carranza, E., & Hale, M., (2005). Arsenic geochemistry and health. *Environ. Int., 31*, 631–641.
3. World Health Organization, (2004). *Compendium of Indicators for Monitoring and Evaluating National Tuberculosis Programs.* No. WHO/HTM/TB/2004.344. Geneva: World Health Organization.
4. Gude, J. C. J., Rietveld, L. C., & Van, H. D., (2016). Fate of low arsenic concentrations during full-scale aeration and rapid filtration. *Water Res., 88,* 566–574.

5. Singh, R., Singh, S., Parihar, P., Singh, V. P., & Prasad, S. M., (2015). Arsenic contamination, consequences, and remediation techniques: A review. *Ecotoxicol. Environ. Saf., 112*, 247–270.

6. Choong, T. S. Y., Chuah, T. G., Robih, Y., Koay, F. L. G., & Azni, I., (2007). Arsenic toxicity, health hazards, and removal techniques from water: An overview. *Desalination, 217*, 139–166.

7. Smedley, P. L., & Kinniburgh, D. G., (2002). A review of the source, behavior, and distribution of arsenic in natural waters. *Appl. Geochem., 17*, 517–568.

8. Al Rmalli, C. F., Harrington, M., Ayub, P. I., & Haris, A., (2005). Biomaterial based approach for arsenic removal from water. *J. Environ. Monit., 7*, 279–282.

9. Akin, I., Arslan, G., Tor, A., Ersoz, M., & Cengeloglu, Y., (2012). Arsenic (V) removal from underground water by magnetic nanoparticles synthesized from waste red mud. *J. Hazard. Mater., 235*, 62–68.

10. Gandhi, M. R., Viswanathan, N., & Meenakshi, S., (2010). Preparation and application of alumina/chitosan biocomposite. *Int. J. Biol. Macromolec., 47*, 146–154.

11. [Subbaiah, M. V., Yuvaraja, G., Vijaya, Y., & Krishnaiah, A., (2011). Equilibrium, kinetic, and thermodynamic studies on biosorption of Pb (II) and Cd (II) from aqueous solution by fungus (*Trametes versicolor*) biomass. *J. Taiwan. Inst. Chem. Eng., 42*, 965–971.

12. Gupta, A., Yunus, M., & Sankararamakrishnan, N., (2012). Zerovalent iron encapsulated chitosan nanospheres: A novel adsorbent for the removal of total inorganic Arsenic from aqueous systems, *Chemosphere, 86*, 150–155.

13. Sashiwa, H., & Aiba, S. I., (2004). Chemically modified chitin and chitosan as biomaterials. *Prog. Polym. Sci., 29*, 887–908.

14. Duan, J., Han, C., Liu, L., Jiang, J., Li, J., Li, Y., & Guan, C., (2015). Binding cellulose and chitosan via intermolecular inclusion interaction: Synthesis and characterization of gel. *J. Spectrosc., 2015*.

15. Elfeky, S. A., Imam, H., & Alsherbini, A. A., (2013). Bio-absorption of Ni and Cd on *Eichhornia crassipes* root thin film. *Environ. Sci. Pollut. Res., 20*, 8220–8226.

16. Verma, V. K., Tewari, S., & Rai, J. P. N., (2008). Ion exchange during heavy metal bio-sorption from aqueous solution by dried biomass of macrophytes. *Bioresource Technol., 99*, 1932–1938.

17. Bhatt, A. S., Sakaria, P. L., Vasudevan, M., Pawar, R. R., Sudheesh, N., Bajaj, H. C., & Mody, H. M., (2012). Adsorption of an anionic dye from aqueous medium by organoclays: Equilibrium modeling, kinetic, and thermodynamic exploration. *RSC Adv., 2*, 8663–8671.

18. Ho, Y. S., (2006). Isotherms for the sorption of lead onto peat: Comparison of linear and non-linear methods. *Pol. J. Environ. Stud., 15*, 81–86.

19. Southichak, B., Nakano, K., Nomura, M., Chiba, N., & Nishimura, O., (2006). Phragmites austral is: A novel bio-sorbent for the removal of heavy metals from aqueous solution. *Water Res., 40*, 2295–2302.

20. Delley, B., (2000). From molecules to solids with the DMol³ approach. *J. Chem. Phys., 113*, 7756–7764.

21. Delley, B., (1990). An all-electron numerical method for solving the local density functional for polyatomic molecules. *J. Chem. Phys., 92*, 508–517.

22. Perdew, J. P., Burke, K., & Wang, Y., (1996). Generalized gradient approximation for the exchange-correlation hole of a many-electron system. *Phys. Rev. B, 4,* 1653316539.

23. Appalakondaiah, S., Vaitheeswaran, G., Lebegue, S., Christensen, N. E., & Svane, A., (2012). Effect of van der Waals interactions on the structural and elastic properties of black phosphorus. *Phys. Rev. B, 86,* 035105–035113.

24. Ortmann, F., Bechstedt, F., & Schmidt, W. G., (2006). Semi-empirical van der Waals correction to the density functional description of solids and molecular structures. *Phys. Rev. B, 73,* 205101–205111.

25. Zhou, J., Xiao, F., Wang, W. N., & Fan, K. N., (2007). Theoretical study of the interaction of nitric oxide with small neutral and charged silver clusters. *J. Mol. Struct., 818,* 51–55.

26. Iman, M., Bania, K. K., & Maji, T. K., (2013). Green jute-based cross-linked soy flour nanocomposites reinforced with cellulose whiskers and nanoclay. *Ind. Eng. Chem. Res., 52,* 6969–6983.

27. Zheng, J. C., Feng, H. M., Lam, M. H. W., Lam, P. K. S., Ding, Y. W., & Yu, H. Q., (2009). Removal of Cu (II) in aqueous media by bio-sorption using water hyacinth roots as a bio-sorbent material. *J. Hazard. Mater., 171,* 780–785.

28. Nayak, P. S., & Singh, B. K., (2007). Instrumental characterization of clay by XRF, XRD, and FTIR. *Bull. Mater. Sci., 30,* 235–238.

29. Cordeiro, N., Gouveia, C., & Jacob, J. M., (2011). Investigation of surface properties of physico-chemically modified natural fibers using inverse gas chromatography. *Ind. Crops. Prod., 33,* 108–115.

30. http://proteopedia.org/wiki/index.php/Hydrogen_bonds; https://upload.wikimedia. org/wikipedia/commons/3/35/WikipediaHDonorAcceptor.png (accessed on 8 August 2020).

31. Prashanth, K. H., & Tharanathan, R. N., (2007). Chitin/chitosan: Modifications and their unlimited application potential-an overview. *Food Sci. Technol., 18,* 117–131.

32. Nishiyama, Y., Langan, P., & Chanzy, H., (2002). Crystal structure and hydrogen-bonding system in cellulose Iβ from synchrotron x-ray and neutron fiber diffraction. *J. Am. Chem. Soc., 124,* 9074–9082.

33. Cai, Z., & Kim, J., (2008). Characterization and electromechanical performance of cellulose-chitosan blend electro-active paper. ʃ *Smart Mater. Struct., 17,* 035028–035035.

34. Kumar, S., & Koh, J., (2012). Physiochemical, optical, and biological activity of chitosan-chromone derivative for biomedical applications. *Int. J. Mol. Sci., 13,* 6102–6116.

35. Srivastava, S., Agrawal, S. B., & Mondal, M. K., (2015). A review on progress of heavy metal removal using adsorbents of microbial and plant origin. *Environ. Sci. Pollut. Res., 22,* 15386–15415.

CHAPTER 5

Computational Investigations on Metal Oxide Clusters and Graphene-Based Nanomaterials for Heterogeneous Catalysis

NEETU GOEL, NAVJOT KAUR, MOHD RIYAZ, and SARITA YADAV

Theoretical and Computational Chemistry Group,
Department of Chemistry and Center of Advanced Studies in Chemistry,
Panjab University, Chandigarh – 160014, India,
E-mail: neetugoel@pu.ac.in (N. Goel)

ABSTRACT

Over the past few decades, nanomaterials have immensely enticed the scientists owing to their extraordinary properties. Association of atoms at nanoscale level constitutes a bridge between single molecules and infinite bulk systems and imparts them unique characteristics. In the recent years, nanoporous solids and clusters have attracted great interest owing to their exceptional properties and wide applicability as both catalysts and catalytic supports in industrial processes. Advances in scientific approach for catalyst design and developing novel functional materials through two molecular aggregations have generated interest among researchers to scrutinize the processes at atomic level. Future development in nanotechnology significantly depends on the fundamental understandings of structure and dynamics of nanomaterials that can be provided by multi-scale modeling and simulation. Theoretical techniques and mathematical modeling efforts are now well advanced to make re-liable predictions about the properties of materials to promote their application in electronics, photonics, catalysis, sensor, and energy storage via size scaling and structural modification.

This chapter presents significant contributions of computational studies in design and development of heterogeneous catalysts based on graphene and transition metal oxides.

5.1   INTRODUCTION

Civilization has traversed a path of confound development with rapid strides in the 21st century owing to fast and cheap production of diverse new exciting materials in large quantities. This advancement could be accomplished with the implementation of heterogeneous catalysis that has provided new vistas for the affordable production of desired materials. Catalysis is a major realm of chemistry; it is directly or indirectly involved not only in laboratory experiments but also in large scale industrial processes. Catalyst has huge impact on several manufacturing processes such as those of medicines, chemical, and paints, etc., as it accelerates the reaction rate by lowering the activation barrier. The design of catalyst needs crucial consideration while planning chemical synthetic procedures. It requires complete understanding of the mechanism involved in the reaction and the knowledge of vital structural parameters that govern activity, selectivity, and lifetime of catalyst and interdependence of these vital factors. Origin of catalyst design is based on the interrelation between structure and properties of materials (structure-activity relationships). Tailoring of textural properties, i.e., geometrical shape, surface area, and pore structure is an important aspect of catalyst design. It is a multidisciplinary venture, which is located at the interface of chemistry, chemical engineering, and material science.

Traditionally, catalysts have been classified as homogeneous and heterogeneous. In homogeneous catalyst, molecules are in same phase as reactant while the phase of molecules in heterogeneous catalyst is different from that of reactant. Though homogeneous catalyst demonstrates exceptional catalytic behavior along with high selectivity and efficiency at milder temperature and pressure, however, its separation from reaction mixture is intricate. Advent of novel materials with potential catalytic properties and ease of their removal has generated lot of interest in heterogeneous catalysis. Design and development of new catalysts to boost the production of essential commodities is highly desirable to cater to the demands of modern society. In order to achieve this, it is necessary to replace present technologies with new, energy-efficient, environment-friendly catalytic processes of high selectivity. This quest has triggered a renewed interest in metal oxide and carbon-based nanomaterials owing to their large surface area, chemical stability, and remarkable catalytic properties.

Among different carbon-based materials, graphene is considered as the basic building block of materials of all dimensions. It consists of single atomic layer of sp^2 bonded carbon atoms in honeycomb crystal lattice structure. This unique nanostructure promises potential applications in technological fields such as catalysis [1, 2], optical electronics [3], sensors [4, 5], and energy conversion [6, 7], etc. Since graphene is semi-metallic in nature with zero band gap and low density of state at the Fermi level, the modifications of this unique nanosurface present a gateway to tap it for applications in catalysis [8]. The electronic state of the 2D graphene crystal lattice can be controlled by several ways that include by introducing defects, doping, and functionalization with hetero-atoms. Sandwiching nano-reactor between the crystals provides increased flexibility in regulating the chemical reactivity of such systems [9, 10]. In addition, edge effects also impart enhanced electronic density of state at the Fermi level on the edges of graphene (zigzag and armchair) and can be useful in catalytic applications [11].

Like modified graphene, metal oxides also hold much significance in the science of catalyst design. Principally, the coordinative unsaturation in metal oxides is the key to their catalytic activity. Metal oxides are made up of cations and anions; variable valency of cation enables it to undergo simultaneous oxidation and reduction. In addition to being used as catalysts, metal oxides are also precursors for other important catalysts. The foremost use of metal oxides in catalysis is towards oxidation reactions, on which a large number of industrial processes are based. The reactivity of oxygen in case of metal oxide is strongly dependent on the kind of neighboring metal ions as well as M-O bonding distance and bond strength.

Clusters with sizes of few nanometers show the highest catalytic efficiencies, as the number of surface sites is high. They are often produced in gas phase but will always be deposited on some sort of surface such as silica, alumina, silica-alumina, carbons, zeolites, mesoporous silicas as MCM41, SBA-15, or metal-organic frameworks (MOFs) for further analysis and applications. The solid support not only anchors the cluster, the synergistic interaction between the cluster and the support surface leads to increase in electron and thermal conductivity of the cluster species and enhances their resistance toward degradation in acid and alkali media. Metal oxides grafted on graphene surface exhibit high stability, activity, and selectivity and have been extensively investigated. Owing to incredible progress in speed and accuracy of theoretical methodologies and computing power, computational simulations and modeling have become reliably predictive. Subsequently, computational chemistry has come to be recognized as an important tool in the arsenal of chemists, specially in the field of catalysis. Sophisticated quantum mechanical (QM) methods have been designed that can be used to quantify the energetics and properties while designing novel materials. Computational simulations provide information that is beyond the realm of experiments, for example, we can exploit properties of short-lived intermediates formed during the course of a re-action which are very difficult to detect and characterize experimentally. It is beyond doubt that computational simulations will make immense contribution in the field of material design. The present chapter focuses on computational advances made so far and the challenges ahead in designing heterogeneous catalyst based on graphene and metal oxide clusters.

5.2   GRAPHENE-BASED NANO MATERIAL AS CATALYST

Although pristine graphene lacks catalytic activity due to low density of states at the Fermi level, yet there are reports which suggest the possibility of using it as catalysts in organometallic complexation reactions that make use of its delocalized π-electron system [12]. The suitability of a catalyst is analyzed by the nature, concentration, and capability of the active sites to effectively adsorb the reactants and stabilizing the surface intermediates. It has been established by numerous literature reports that catalytic efficiency of pristine graphene is significantly enhanced by its doping or functionalization.

The heteroatom doped graphene is considered as one of the most promising metal free catalysts. There are mainly two types of heteroatom-doped catalysts, i.e., noble-metal-free (transition metal-doped graphene) and metal-free catalysts (doping with B, N, S, P, etc.). In metal-doped graphene, generally metal is the active site, while in metal-free catalysts, the doped heteroatom induces redistribution of charge density that disturbs the uniform charge cloud of graphene making it favorable for adsorption of reactants. Functionalized graphene such as graphene oxide (GO) has also been deemed as proficient catalyst, it is an analog of graphene having 2D structure with many oxygen-containing functional groups (i.e., epoxy, hydroxyl, or carboxylic) [13] that are responsible for its high catalytic activity in a wide range of chemical reactions.

It has been shown that hydroxyl groups on GO surface possess bi-functional effects that enhance adsorption of SO_2 through H-bonding interaction and reducing the reaction barrier for its oxidation to SO_3 [14]. Based on Bader population, charge difference, and electron localization function analysis, a charge transfer channel has been proposed to explain enhanced rate of oxidation of SO_2 to SO_3. Bielawski and co-workers [15] have revealed the potential of harnessing the reactivity of GO for various synthetic reactions. Capitalizing on the unique chemistry inherent to GO, the authors demonstrated the efficient oxidation of benzyl alcohol to benzaldehyde (conversion > 90%) in the presence of GO as a hetero-geneous catalyst. Over oxidation to benzoic acid was observed in only minimal amounts (7%) and only under certain conditions (e.g., at elevated temperatures). Interestingly, this, and other oxidation reactions of alcohols were performed under ambient conditions and did not proceed under a nitrogen-blanketed atmosphere, which suggests that oxygen may be func-tioning as the terminal oxidant. This study is a compelling demonstration of the new concept of using large-area (metal-free) GO as catalysts, aptly coined "carbo catalysis" by the authors. Recent studies have also indicated that decoration of GO with metals such as Ti [16], Li, Al [17], or Pd [18] could enhance the surface reactivity to separate target gases significantly. Interaction between the decorated metals is usually found to be weaker as compared to the interactions between the metal atoms and GO, resulting in non-aggregation of metal atoms to form clusters. For instance, Lv et al. [19] have claimed that the adsorption and decomposition of N_2O molecule could be easily performed over Al-decorated GO. Esrafili et al. have reported that the Si-decorated GO could be regarded as an active metal-free catalyst for the oxidation of CO [20] and reduction of N_2O [21].

The ease and accuracy of computational modeling of graphene-based nano-materials has created opportunities to design and develop novel catalysts for challenging reactions like CO oxidation, Oxygen reduction reaction (ORR), Oxygen evolution reaction (OER) and Nitrogen reduc-tion reaction (NRR), etc., [22–24] using modified graphene. The reactions like ORR and OER have attracted extensive interest due to their merit for producing clean and sustainable form of energy. NRR is also an important and challenging chemical process which is critical for the production of ammonia. A facile and efficient catalysts for above reactions are necessary to withstand environmental pollution, reduce waste product and for a sustainable human development. The subsequent section will dwell upon computational advances in designing graphene-based efficient catalysts for these chemical processes and elucidation of their reaction mechanisms. Each reaction will be briefly discussed to analyze the catalytic efficiency of employed graphene-based material and further scope of development.

5.2.1 CO OXIDATION

To regulate the emission of CO from automobile industry, its oxidation is one of the most important reactions. The CO oxidation reaction is considered as a benchmark for heterogeneous catalysis [25], it is an exothermic process, but the reaction is spin-forbidden in gas phase as the reactant (O_2) has triplet ground state while the total spin of the product (CO_2) is zero. Traditionally, noble metals like Pt, Pd, and Au serve as catalysts for CO oxidation [26–29]. However, their high cost and high-temperature requirement have instigated the need for better catalysts. Recent studies have focused on reducing noble metal content of the catalysts and explore transition metal like Cr, Mn, Fe, Co, Ni, and Cu based catalysts by combining them with organic/inorganic polymers to increase their stability and catalytic activity [30–32]. In the quest of designing efficient catalysts for CO oxidation, it was observed that graphene in conjugation with transition metal exhibits promising potential as catalysts for CO oxidation [33]. Li et al. [23] have explored CO oxidation using Fe embedded graphene-metal system by density functional computations. Strong hybridization between 3d orbital of the embedded Fe and 2p orbital of O_2 molecule efficiently activates the O_2 molecule, making insertion of CO between O=O bond easy with activation barrier of only 0.58 eV. It has been illustrated through theoretical and experimental investigations that co-doping of graphene with metal and non-metal shows synergistic effect to enhance the catalytic activity [34]. In this context, Tang et al. [35] considered co-doping graphene with non-metal (N, Si, P) for CO oxidation at low temperature. The authors report that electro negativity difference between the co-doped atoms can be used to control the charge over the dopant atom which helps in regulating the stability of the reactant over co-doped graphene surface.

It has been demonstrated that CoN_3 doped graphene is the most efficient catalyst for CO oxidation with energy barrier less than 0.5 eV. We directed our research endeavors to investigate catalytic efficiency of pristine graphene, epoxy functionalized and sulfur-doped graphene towards CO oxidation [36]. Our DFT computations for the adsorption of the reactant species, i.e., CO, and O_2 molecules over graphene surfaces suggested strong adsorption of O_2 over sulfur-doped divacant graphene (S2-Gra) with adsorption energy of −17.29 kcal/mol. Charge transfer from S-doped graphene surface to the anti-bonding orbital of O_2 molecule.

Leads to activation of O_2 molecule. The mechanistic pathway for CO oxidation was explored through well-established Langmuir-Hinshelwood (LH) and the Eley-Rideal (ER) mechanisms (Figure 5.1). The LH mechanism proceeds by reaction between co-adsorbed O_2 and CO molecules over the surface while ER mechanism involves reaction of CO with pre-adsorbed O_2 to form the product. Our density functional calculations suggested that adsorption energy of O_2 over S-doped graphene surfaces is higher than CO, thus the likelihood of co-adsorption of CO and O_2 was ruled out. CO oxidation was envisaged to proceed through ER-mechanism that involved a carbonate intermediate formed by the insertion of one CO between O=O bond via a transition state. The 2nd pre-adsorbed CO then interacts with the carbonate intermediate leading to the formation two CO_2 molecules. S-doped graphene surface effectively captures and activates O_2 molecule and directs CO oxidation through low activation barrier (of only 0.61 eV).

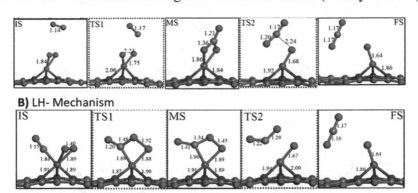

FIGURE 5.1 Schematic depiction of the two reaction mechanisms: (A) ER-Mechanism and (B) LH-Mechanism for CO oxidation over metal Co-doped graphene surface Ref. [37] (IS = initial state, MS = metastable state and TS = transition state).

Source: Reprinted with permission Ref. [37] 2016. © Elsevier.

5.2.2 OXYGEN REDUCTION/EVOLUTION REACTION ORR/OER

ORR is the key process in fuel cell to convert chemical energy to electrical energy. It is a sluggish process that occurs at the cathode of the cell and decides efficiency of the cell. Pt based catalysts are the most common for ORR despite the limitation of their high cost and CO poisoning [38]. Extensive research has been devoted towards the development of economic and environment-friendly catalysts for ORR. Recent theoretical and experimental studies [39, 40] suggest graphene-based catalysts as active and cost-effective alternative to Pt-based catalysts for ORR. Two mechanisms have been proposed for the ORR (Figure 5.2), one is 4e reduction pathway in which there is direct reduction of O_2 to H_2O, and other is 2e reduction pathway in which O_2 is first reduced to H_2O_2 and then further reduced to H_2O. Various research groups have shown N-doped graphene with hetero-atoms as efficient catalysts for ORR in alkaline conditions

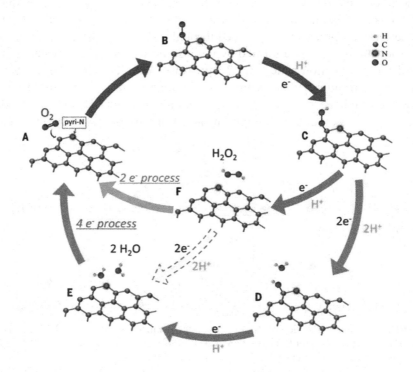

FIGURE 5.2   Schematic pathway for ORR on N-doped graphene Ref. [43].

due to its high specific area. Zhang et al. [41] have demonstrated N-doped graphene as excellent electrocatalysts for ORR. High asymmetric spin density and atomic charge density induced by nitrogen doping is shown to be responsible for high electron catalytic activity. In addition to Nitrogen, other metals and non-metal doped graphene have also been reported as excellent electro catalysts for ORR. For instance, computations by Sun et al. [42] have demonstrated FeN_4 and $Fe(CN)N_4$ embedded models of graphene as efficient catalysts for ORR. They concluded that both 4e and 2e reduction pathways involve the formation of OOH intermediate. The energetics showed a penta-coordinated $Fe(CN)N_4$ embedded graphene model as efficient catalysts for ORR with a calculated barrier of only 0.40 eV at an applied potential of 0.8 V which is 0.2 eV lower than those determined over Pt(111) catalyst.

Reverse reaction of ORR, i.e., OER is the key process for water splitting which is the blueprint for the realization of an effective and efficient hydrogen economy. The OER is sluggish due to the requirement of four-electron transfer in the process with substantial over potential. Traditionally, precious metals like Pt, Ru, Ir, and their oxides serve as catalysts for OER [39, 44, 45]. It is necessary for the widespread utilization of the hydrogen economy to substitute these metal oxides with non-precious elements. Recent studies in this regard suggested that graphene plays an important role in OER via electron catalytic reaction. The proposed reaction mechanism for OER involves the formation of intermediates like M-OH, M-O, M-OOH (M=surface) and the stability of these intermediate and the different approaches to form O_2 molecules from these intermediates are crucial for the electrocatalytic activity [46]. Over potential is an important descriptor for designing catalysts for OER, it arises from the energy difference between the involved intermediates. The ORR mechanism can be summarized in the following four elementary steps:

$$H_2O + M \rightarrow M - OH + (H^+ + e^-) \tag{1}$$

$$M - OH \rightarrow M - O + (H^+ + e^-) \tag{2}$$

$$M - O + H_2O \rightarrow M - OOH + (H^+ + e^-) \tag{3}$$

$$M - OOH \rightarrow M + O_2 + (H^+ + e^-) \tag{4}$$

The free energy change for the overall conversion process is 4.92 eV. So an ideal catalyst should be able to felicitate all the four charge transfer steps with same reaction free energies (i.e., 4.94/4 = 1.23 eV) and its equivalent to being zero reaction free energy at equilibrium potential, i.e., 1.23 V. In the quest of minimizing the overpotential for OER, Mao et al. [47] investigated single and double vacancy graphene supported over Ni and Co metal surface as catalyst for OER. Through DFT computations, the author showed that the integrated structure can be designed as a bi-functional catalyst for OER and ORR. The double vacant graphene over Co(0001) surface acts as an excellent catalyst for OER and ORR with small overpotential of only 0.39 V for OER and 0.36 V for ORR. Nitrogen-doped armchair nanoribbons of graphene are reported to work as excellent catalysts for OER with an overpotential of only 0.40 V by Li et al. [48]. In another study, Guan et al. [49] have explored both theoretically and experimentally the activity of mononuclear manganese embedded in N-doped graphene as catalysts for OER. It was observed that MnN_4 embedded graphene exhibits better catalytic activity than Mn embedded graphene owing to the four N-coordinated Mn that favor the formation of Mn^{IV}-O species and felicitates the nucleophilic attack of the second H_2O to Mn-O species to give O_2 molecule via Mn-OOH intermediate. Computational studies have immensely contributed to the field of heterogeneous catalysis by providing mechanistic details of the reaction process at atomic scale and thus act as guiding light for the experimental efforts.

5.2.3   NITROGEN REDUCTION REACTION (NRR)

The production of ammonia from atmospheric nitrogen, i.e., NRR is another key process of vital importance. Due to the strong non-polar bonding of N_2, NRR is a huge challenge and electrochemical conversion of N_2 to NH_3 provides a sustainable alternative for NRR. Primarily, Haber-Bosch process is used for transformation of N_2 to ammonia in which iron-based catalyst is used under very harsh conditions and the process consumes a huge amount of energy. The development of an alternative to the Harber-Bosch process is highly desirable. Li et al. [50] have investigated the suitability of FeN_3-embedded graphene as a catalyst for NRR using first principle approach. Their calculation suggests that the graphene model embedded with FeN_3 effectively captures and activates the inert gaseous N_2. The incorporation of three N atoms with Fe in the graphene model drastically enhanced the spin moment of Fe and induced a spin-polarized ground state. The system remains heavily spin-polarized even after the adsorption of N_2 which is responsible for the easy hydrogenation of the adsorbed N_2 at low temperature. We relied on DFT computations to design a catalyst for NRR, embedded in divacant-graphene (DVG) [51]. The DVG was embedded with different transition metals and vacant sites were passivated with four nitrogen atoms. It is pertinent to mention here that the doping of transition metals in DVG has been achieved experimentally. Out of Cr, Mn, Fe, Mo, Ru embedded in DVG, Cr shows the highest affinity for gaseous N_2 with an adsorption energy of -0.65 eV. The NRR over the surface can follow either the Distal (for the end-on adsorption configuration) or the enzymatic pathway (for side-on adsorption configuration; Figure 5.3). The conversion process consists of six consecutive protonation and reduction steps. Energy profile for each step of the enzymatic pathway including the energy for desorption of NH_3 from the surface is given in Figure 5.4. All the steps were exothermic in nature except the desorption of second NH_3 species from the substrate which is endothermic by 2.18 eV. Desorption step can be facilitated by further protonation of NH_3 to NH_4^+, it brings down the barrier to 0.71 eV, and NH_4^+ ion Obtained as the reduced product can be easily harvested using a suitable counterion. This is concordant with experimental reports that obtained NH_4^+ and $N_2H_5^+$ as products of N_2 reduction. Attention has also been paid towards developing non-transition metal catalysts for NRR in ambient condition using DFT computations, Tian et al. [52] have demonstrated Al-doped graphene as an efficient catalyst for NRR, the embedded Al acts as active site for binding N_2 molecule while graphene framework acts as redox center (charge buffer) during the protonation and reduction steps. The authors also used Li^+ ion as an additive which felicitates the transfer of hydrogen and exchange of NH_3 with N_2 for the next catalytic cycle. The exchange of NH_3 with N_2 is the last step in the catalytic cycle and also the rate-determining step with an energy barrier of only 0.80 eV. In light of the above works, it's important to note that an efficient catalyst for NRR should be of good electrical conductivity, should bind with N_2 and the intermediates moderately and should be a good redox agent.

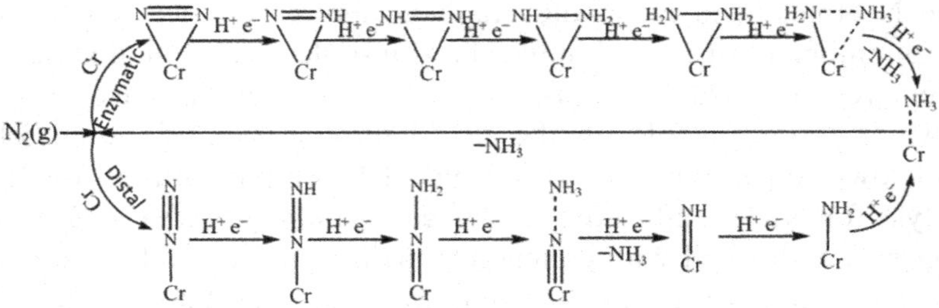

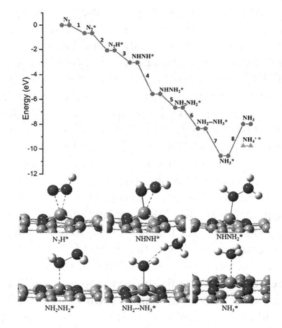

FIGURE 5.3 Schematic depiction of the two reaction mechanisms for NRR over GraN4-Cr system [51].

Source: Reprinted with permission from Ref. [51]. © John Wiley.

FIGURE 5.4 Energy profile for the NRR over GraN4-Cr and optimized geometries of various intermediates along the reaction path through enzymatic mechanism [51].

Source: Reprinted with permission from Ref. [51]. © John Wiley.

5.3   METAL OXIDE CLUSTER AS CATALYST

Among all catalysts used at the industrial scale, metal oxides hold high significance and are essential in almost all refining and petrochemical processes. Metal oxides gained prominence in the mid-1950s when they were found to effectively catalyze a wide variety of reactions, in particular oxidation and acid-base reactions. Due to large surface fraction, interest continues to grow towards the use of gas-phase clusters in order to model reactions occurring over surface of metal oxides. The employment of gas-phase cluster as models of heterogeneous catalysts has been well established in literature. It is an ideal approach to obtain molecular-level understanding of catalytic active sites and reaction mechanisms. Numerous studies have encouraged that a metal oxide surface may be viewed, to a first approximation, as a collection of clusters [53]. The similarities between the electronic structure of these "cluster-like" surface species and that of gas-phase clusters [53] allow information about potential active sites to be gained from the study of gas-phase cluster ions and neutrals [54].

Heterogeneous catalysis using metal oxides is an important solution to abate environmental pollution. The increase in the emission of harmful gases such as nitrogen oxides (NO_x) and sulfur dioxide (SO_2), from the atmosphere not only pose a threat to health, but also cause degradation of buildings [55]. Since, NO_2 (Nitrogen dioxide) can be easily removed than NO by absorption, efforts have therefore been taken to develop catalysts for oxidation of NO to NO_2 [41, 56]. Various metal oxide clusters with supported catalysts have been used for NO oxidation [57, 58]. Recently, Zhang et al. [41] have reported the catalytic oxidation of NO with O_2 over $FeMnO_x/TiO_2$, the mixed Mn-Ce oxide is widely used for NO oxidation [56]. The use of cobalt oxide cluster for NO oxidation has been studied by Xie et al. [59]. We investigated the possibility of NO oxidation using $(CrO_3)_3$ cluster in its six-membered nonplanar ring with chair conformation [60]. The cluster had terminal as well as bridged oxygen's; the bridged oxygen did not show any affinity for NO molecule while NO_2 and Cr_3O_8 are obtained as products by oxidation of NO by the terminal oxygen atoms of the cluster. The low Transition State (TS) barrier and negative Gibbs free energy change for all the oxidation steps confirmed the suitability of Cr_3O_9 cluster as a potential oxidant to convert NO to NO_2.

Most prevailing research efforts in catalytic chemistry are to develop methods for activation of C-C and C-H bonds of organic molecules and oxidation of hydrocarbons. It has been reported that efficient C-H bond activation can take place over many atoms, ions, and atomic clusters, among which TMO clusters [61–63] are an important type. Unraveling the mechanisms of these reactions is one of the most challenging projects in chemistry and continues to be an intensively studied subject. It has been shown by DFT studies that cluster containing oxygen atoms with unpaired spin density (UPSD) act as oxygen-centered radicals (O$^-$). The O$^-$ radicals over the studied TMO clusters can effectively abstract hydrogen atoms from the reacting alkane molecules. DFT calculations have been performed to study the structure and reactivity of small clusters $(Sc_2O_3)_{1-3}^+$. The experimentally observed C-H bond activation by $(Sc_2O_3)_N^+$ is facilitated by oxygen-centered radicals bonded as a bridge in the clusters [64]. TMO's clusters containing the O$^-$ moiety have generated promising results for low-temperature C-H bond activation [65]. Most early studies of methane activation by oxide clusters focused on the TMO clusters with character of oxygen-centered radicals such as FeO$^+$ [66], MoO$_3^+$ [67], OsO$_4^+$ [67] and V$_4$O$_{10}^+$ [63, 68]. Fiedler et al. [69] have demonstrated that CrO$_2^+$ is able to activate H-H, C-H, and C-C bonds in different hydrocarbons. The reactivity of charged $V_xO_y^{+/-}$ clusters has been extensively investigated towards activation of hydrocarbons viz. methane [70], ethane [71], propane [72], 1,3-butadiene (C$_4$H$_6$), deuterated 1,3-butadienes C$_4$D$_6$ and 1,1,4,4-C$_4$D$_4$H$_2$ [73]. Dong et al. [74] have demonstrated that neutral V$_x$O$_y$ clusters display a considerably different reactivity than the $V_xO_y^{+/-}$ clusters towards ethane. Aldehydes [73] and association products [75] have been reported with V$_x$O$_y$ for C-H bond activation of aliphatic hydrocarbons. We present here the oxidation of aliphatic and aromatic hydrocarbons aided by chromium and vanadium oxide.

5.3.1   OXIDATION OF ALIPHATIC HYDROCARBONS

Chromium belongs to the first-row transition metal series and exhibits high spin 7S$_3$ ([Ar]3d^{5}4s^1) ground state electronic configuration. Oxides of Chromium hold an important position in a number of industrial processes as catalyst due to their ability to show variety of oxidation states and various degree of polymerization [76–78]. Consequently, they can form wide range of bulk oxide such as CrO$_2$, Cr$_2$O$_3$ and Cr$_5$O$_{12}$ [79]. Out of which Cr$_2$O$_3$ is the most stable with metal atoms occupying two-thirds of the octahedral sites between two layers [80]. The charge transfer between Cr and O in chromium oxides plays an important role in defining the spin on the metal, which affects their magnetic and catalytic activity. This exciting feature offers opportunities to tailor the chemical and physical properties of chromium oxides for various applications. We investigated the mechanism of ethane, ethene, and ethylene activation by CrO_3 cluster. The reaction of metal oxide with hydrocarbon molecules may activate either the C-H or the C-C bond of the hydrocarbon. Both these possibilities were explored by considering different relative orientations of the reacting molecules to ensure that the metal oxide has a chance either to react with the hydrogen or to attack the C-C bond [81]. It has been observed that the final products are sensitive to the relative configurations of the two reactants shown in Figure 5.5. The placement of CrO_3 cluster near the C-H bond of ethane leads to the formation of ethanol with the liberation of 0.17 kcal/mol of energy. More than one

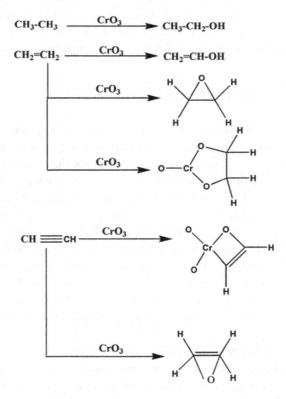

FIGURE 5.5 Different products encountered in the reaction of aliphatic hydrocarbons with CrO_3.

product are formed in case of ethane, as both C-C and C-H bond gets activated by CrO_3 cluster unit. The activation of C-H bond leads to the formation of ethanol while C=C bond results in the formation of two products, i.e., epoxide, and association product. Similarly, the reactivity of CrO_3 cluster towards the activation of C-H and C-C bonds of ethane has also been investigated in detail. It was observed that CrO_3 clusters can activate only the C-C triple bond of ethane. Activation of both C-H and C-C bond of acetylene is reported for its reaction with vanadium oxide clusters [82], product formed via C-C triple bond activation is found to be thermodynamically favorable at room temperature. In the case of reaction with neutral cobalt oxide clusters [59], the reported product results from C-H bond activation of acetylene.

Numerous reports on activation of C-H/C-C bonds of aliphatic hydrocarbons are present in literature but studies on the activation of aromatic hydrocarbons are very less. The deformation of π-electron cloud resulting into loss of resonance energy is the main challenge in this process.

5.3.2   OXIDATION OF BENZENE

The activation and hydroxylation of benzene leads to the formation of Phenol, which is an important intermediate used in various syntheses like petrochemicals, agrochemicals, and plastics [83]. Presently, cumene process is used for the production of phenol from benzene but apart from the multi-step process, production of equimolar amounts of acetone as a byproduct is a main shortcoming of this process [84]. Yoshizawa and coworkers have suggested direct hydroxylation of benzene with the activation of C-H bond of aromatic ring by FeO^+ [85]. Different experimental and theoretical studies on iron-modified zeolites Fe-ZSM-5 for the highly selective gas phase oxidation of benzene by N_2O have been performed. The unique oxidation property of Fe-ZSM-5 catalyst is owing to the extra framework oxygen (α-oxygen) formed by the decomposition of N_2O [86].

Oxidation of benzene to phenol using cationic oxide clusters in the gas phase has been demonstrated in many studies that cover experimental as well as theoretical aspects of the reaction [85–90]. Shiota et al. [88] have studied the direct benzene-phenol oxidation using cationic iron oxide cluster and out of the three proposed mechanistic pathways, i.e., non-radical, radical, and oxygen-insertion mechanism, the non-radical mechanism is reported to be the favored one. Chromium oxide clusters are known to be more reactive and selective in oxidation reactions than early or late TMO clusters. It was evident in the reaction of CrO_3 with benzene that led to two products viz. phenol and benzoxide Figure 5.6 via C-H and C-C bond activation respectively [81]. Three possible reaction routes for the formation of phenol have been described by Kaur et al. [90], based on the exploration of these mechanistic routes for oxidation of benzene by CrO_3, it is predicted that conversion of benzene into phenol is facilitated via C-H bond activation in the presence of CrO_3. The formation of benzoxide was energetically unstable under ambient conditions of temperature and pressure [81]. Similarly, the reactivity of Vanadium oxide cluster has also been investigated towards benzene, it was observed that the reaction proceeds in single step via three-centered TS [90]. This study provides promising evidence of direct abstraction of hydrogen by terminal oxygen of the cluster via three-centered TS. It is important to note here that low lying excited electronic spin states of TMOs are accessible and they are reported to exhibit different behavior in different spin states in several reactions. We present here two-state reactivity of Vanadium oxide in its reaction with benzene. The different potential energy surface (PES) obtained for the reactivity of the cluster in its singlet state and its triplet analog suggest that stationary points leading to chemical reactivity lie on two PES with the possibility of their nonadiabatic intersection wherein the favorable pathway will not remain on a single PESs (Figure 5.7). The comparison of the reaction profiles for the reactivity of cluster in singlet and triplet states shown in Figure 5.7 reflects that the Spin crossing point lies at the entrance channel and the triplet TS lies lower than the singlet. In this case, the crossing point lies at the entrance channel and allows the reaction to proceed through the lower TS barrier; without the spin-crossover phenomena, it would have required surmounting of a relatively higher barrier. The rigorous computational screening of the reaction between V_4O_{10} cluster and benzene led us to propose an efficient channel for the C-H activation of benzene through the spin crossover. The study elaborates on the direct and selective oxidation of benzene molecule by V_4O_{10} cluster through exhaustive computations performed within the formalism of DFT.

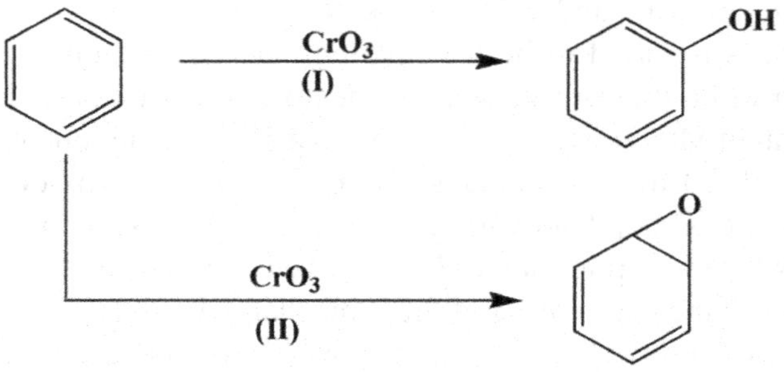

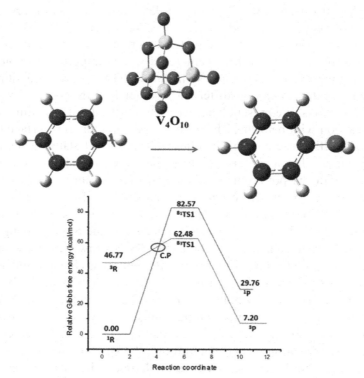

FIGURE 5.6   Possible products for benzene oxidation by CrO$_3$.

FIGURE 5.7   Gibbs free energy profile relative to the reactant in the singlet state (^{1}V$_4$O$_{10}$+C$_6$H$_6$) for the singlet (in black) and triplet state (in red). CP corresponds to the spin crossing point. Relative Gibbs free energies are in kcal/mol [90].

Source: Reprinted with permission from Ref. [90]. © American Chemical Society.

5.4 SUPPORTED METAL OXIDE CLUSTER

Anchoring the cluster catalyst to a substrate is an important aspect in heterogeneous catalysis involving cluster species. The substrate not only holds the cluster but may also take part in catalytic cycle by providing secondary sites for adsorption of reaction intermediates. It is to noted that while gas-phase clusters are the perfect computational model to understand molecular level intricacies of the reaction process, they have little pertinence with the experimental situations. Our QM calculations on size selected metal oxide clusters in gas phase provided sufficient evidence of their engrossing catalytic ability, to mimic experimental conditions reaction mechanisms were reexamined by supporting the cluster on a substrate. The interaction of the cluster with the support not only enhances the cluster stability but may influence its electronic and geometric structure, which in turn opens up new energy relaxation pathways. However, owing to size and complexity of the supported cluster system, it is difficult to perform full QM calculations on these systems. The computational costs for QM methods with increasing size of the systems are poorly scaled; therefore, it is practically impossible to study large systems of supported clusters by rigorous QM methods. To manage these problems, latest development of hybrid theoretical approaches such as ONIOM provides a realistic strategy to study large molecular systems by dividing them into several subsystems (layers) and treating them at different levels of well-developed conventional computational methods rather than employing unrealistic simplified model systems. In the embedded cluster approach, the active site is embedded in a larger cluster of the material, which is usually treated with molecular mechanics (MM) or a relatively cheap QM method and the chemically important region is treated by an expensive and accurate QM method.

The TMOs anchored on a solid support surface are often used in important processes, such as, selective reforming of alkanes, oxidation of alkanes, hydro desulfurization, and other processes [62, 91, 92]. Among all the TMOs, supported chromium species have a rich history in the activation of small molecules and acts as efficient heterogeneous catalytic systems [93–95]. Chromium oxide species supported on silica (CrO_x/ SiO_2) systems, popularly documented as Phillips catalyst are used for ethene polymerization at relatively low pressures [96], dehydrogenation, and dehydroisomerization of alkanes [97, 98], oxidative dehydrogenation of hydrocarbons in the presence of oxygen and CO_2 [99], isomerization, aromatization, complete combustion, selective oxidation reactions [100–103] and methane dehydroaromatization [104]. These catalytic processes are known to happen at some specific active sites on the surface and the reactivity of these depends on the geometric as well as electronic structures of the active sites [105].

To model, the structure of supported chromium oxide, the gas phase CrO_3 cluster was placed on solid silica support $(Si_{12}O_{19}H_{10})$. The current model, i.e., $CrO_3/Si_{12}O_{19}H_{10}$ (abbreviated as CrO_3/SiO_2 hereafter) resembles the stable structure of CrO_x/SiO_2 systems reported in literature [106]. The reactivity of CrO_3/SiO_2 model with ethane and benzene molecules was investigated using ONIOM QM/MM scheme [107]. The reaction system was partitioned into two regions, the bulky silica region was treated with MM, while the chemically important part, i.e., CrO_3 and hydrocarbon molecule, was subjected to QM calculations. Reactivity of CrO_3/SiO_2 towards ethane has been observed by using similar orientations as in gas phase. The detail analysis of mechanism illustrates that in case of CrO_3/SiO_2 ethanol and epoxide are obtained as products. The association product which was obtained in case of gas phase reactions was not obtained here. For the reaction of CrO_3/SiO_2 with C_6H_6 molecule, two major products, phenol, and benzoxide were obtained. In comparison to the reaction with the cluster in gas phase, the phenol formation using CrO_3/SiO_2 species is more exothermic and spontaneous. The computational investigations of gas phase and supported metal oxide clusters summarized here provide reliable understanding of the mechanisms, selectivity, and feasibility of their reaction with aliphatic as well as aromatic hydrocarbons. Such studies provide useful pointers to experimentalists to exploit metal oxides for oxidation processes in the laboratory as well as at industrial scale.

5.5   OUTLOOK AND CHALLENGES

Much fundamental and applied research is devoted to find out how catalysts work and to improve their effectiveness as it has direct impact on cost and energy consumption. The core of designing the metal oxide clusters and graphene-based nanomaterials for heterogeneous catalysis lies in the understanding of the mechanism that sway the reactions and to identify the active sites present over the catalyst. Computational chemistry provides an avenue to unravel the mysteries of the reaction process at the molecular level and thus contribute to the design and development of efficient catalysts. Computational chemistry has been recognized as a vital adjunct to experimental studies; it has become an essential tool just like common laboratory techniques such as nuclear magnetic resonance (NMR) and infrared (IR) spectroscopy. Mostly, computational investigations follow the experimental work and provide a rationale for the obtained results.

This decade has witnessed rapid strides in computational power and the development of nearly accurate electronic structure methods that are beginning to make a deeper impact on the advancement of computational chemistry. It will not only follow the experiments but will guide the experimentalists to plan and design a reaction or a synthesis procedure. Reliable predictions made by computational chemistry will have a huge economic impact as it will reduce the effort in performing tedious, costly, and potentially dangerous experiments. In particular, the level of complexity involved in identifying feasible mechanisms while designing a catalyst is vast. In addition to activity and selectivity, other factors like stability, resistant to surface poisoning, side product formation, and production cost are equally crucial. Computational investigations provide the necessary understanding of these issues in a cheaper and faster manner. It is expected that computational and experimental research communities will work in tandem not only to advance heterogeneous catalysis but also to various frontiers of chemistry.

ACKNOWLEDGMENT

NG is thankful to Dr. Prabhat Ranjan, for the kind invitation to contribute to this book chapter. MR is grateful to University Grants Commission (UGC), New Delhi for senior research fellowship.

KEYWORDS

- **catalysis**
- **graphene**
- **metal oxide**
- **molecular mechanics**
- **oxygen evolution reaction**
- **potential energy surface**

REFERENCES

1. Li, J. C., Hou, P. X., & Liu, C., (2017). *Small*, *13*, 1702002.
2. Li, Y., Fan, X., Qi, J., Ji, J., Wang, S., Zhang, G., & Zhang, F., (2010). *Nano Research*, *3*, 429–437.
3. Nath, P., Sanyal, D., & Jana, D., (2015). *Physica E: Low-Dimensional Systems and Nanostructures*, *69*, 306–315.
4. Gao, L., Lian, C., Zhou, Y., Yan, L., Li, Q., Zhang, C., Chen, L., & Chen, K., (2014). *Biosensors and Bioelectronics*, *60*, 22–29.
5. Leenaerts, O., Partoens, B., & Peeters, F., (2008). *Physical Review B*, *77*, 125416.
6. Yoo, E., Kim, J., Hosono, E., Zhou, H. S., Kudo, T., & Honma, I., (2008). *Nano Letters*, *8*, 2277–2282.
7. Spyrou, K., Gournis, D., & Rudolf, P., (2013). *ECS Journal of Solid State Science and Technology*, *2*, M3160-M3169.
8. Machado, B. F., & Serp, P., (2012). *Catalysis Science and Technology*, *2*, 54–75.
9. Shao, L., Chen, G., Ye, H., Wu, Y., Niu, H., & Zhu, Y., (2014). *EPL (Europhysics Letters)*, *106*, 47003.
10. Tang, C., Wang, H. F., Huang, J. Q., Qian, W., Wei, F., Qiao, S. Z., & Zhang, Q., (2019). *Electrochemical Energy Reviews*, *2*, 332–371.
11. Jeon, I. Y., Choi, H. J., Ju, M. J., Choi, I. T., Lim, K., Ko, J., Kim, H. K., Kim, J. C., Lee, J. J., Shin, D., et al., (2013). *Scientific Reports*, *3*, 2260.
12. Sarkar, S., Niyogi, S., Bekyarova, E., & Haddon, R. C., (2011). *Chemical Science*, *2*, 1326–1333.
13. Shao, Y., Sui, J., Yin, G., & Gao, Y., (2008). *Applied Catalysis B: Environmental*, *79*, 89–99.
14. Zhang, H., Cen, W., Liu, J., Guo, J., Yin, H., & Ning, P., (2015). *Applied Surface Science*, *324*, 61–67.
15. Boukhvalov, D. W., Dreyer, D. R., Bielawski, C. W., & Son, Y. W., (2012). *Chem. Cat. Chem.*, *4*, 1844–1849.
16. Wang, L., Zhao, J., Wang, L., Yan, T., Sun, Y. Y., & Zhang, S. B., (2011). *Physical Chemistry Chemical Physics*, *13*, 21126–21131.
17. Chen, C., Xu, K., Ji, X., Miao, L., & Jiang, J., (2014). *Physical Chemistry Chemical Physics*, *16*, 11031–11036.
18. Tang, S., & Zhu, J., (2014). *RSC Advances*, *4*, 23084–23096.
19. Lv, Z., Mo, H., Chen, C., Ji, X., Xu, K., Miao, L., & Jiang, J., (2015). *RSC Advances*, *5*, 18761–18766.
20. Esrafili, M. D., Sharifi, F., & Nematollahi, P., (2016). *Journal of Molecular Graphics and Modeling*, *69*, 8–16.
21. Esrafili, M. D., Sharifi, F., & Nematollahi, P., (2016). *Applied Surface Science*, *387*, 454–460.
22. Le, Y. Q., Gu, J., & Tian, W. Q., (2014). *Chemical Communications*, *50*, 13319–13322.
23. Li, Y., Zhou, Z., Yu, G., Chen, W., & Chen, Z., (2010). *The Journal of Physical Chemistry C*, *114*, 6250–6254.
24. Chai, G. L., Qiu, K., Qiao, M., Titirici, M. M., Shang, C., & Guo, Z., (2017). *Energy and Environmental Science*, *10*, 1186–1195.

25. Lu, Z., Lv, P., Yang, Z., Li, S., Ma, D., & Wu, R., (2017). *Physical Chemistry Chemical Physics, 19*, 16795–16805.
26. Kummer, J., (1986). *The Journal of Physical Chemistry, 90*, 4747–4752.
27. Manasilp, A., & Gulari, E., (2002). *Applied Catalysis B: Environmental, 37*, 17–25.
28. Grisel, R., & Nieuwenhuys, B., (2001). *Journal of Catalysis, 199*, 48–59.
29. Bekyarova, E., Fornasiero, P., Ka˘spar, J., & Graziani, M., (1998). *Catalysis Today, 45*, 179–183.
30. Szegedi, A., Hegedus, M., Margitfalvi, J. L., & Kiricsi, I., (2005). *Chemical Communications*, 1441–1443.
31. Saalfrank, J. W., & Maier, W. F., (2004). *Angewandte Chemie International Edition, 43*, 2028–2031.
32. Kapteijn, F., Stegenga, S., Dekker, N., Bijsterbosch, J., & Moulijn, J., (1993). *Catalysis Today, 16*, 273–287.
33. Qiu, B., Xing, M., & Zhang, J., (2018). *Chemical Society Reviews, 47*, 2165–2216.
34. Wang, C., Shi, P., Cai, X., Xu, Q., Zhou, X., Zhou, X., Yang, D., Fan, J., Min, Y., Ge, H., et al., (2015). *The Journal of Physical Chemistry C, 120*, 336–344.
35. Tang, Y., Chen, W., Shen, Z., Chang, S., Zhao, M., & Dai, X., (2017). *Carbon, 111*, 448–458.
36. Riyaz, M., Yadav, S., & Goel, N., (2018). *Journal of Molecular Graphics and Modeling, 79*, 27–34.
37. Zhang, X., Lu, Z., & Yang, Z., (2016). *Journal of Molecular Catalysis A: Chemical, 417*, 28–35.
38. Yang, L., Zhao, Y., Chen, S., Wu, Q., Wang, X., & Hu, Z., (2013). *Chinese Journal of Catalysis, 34*, 1986–1991.
39. Suen, N. T., Hung, S. F., Quan, Q., Zhang, N., Xu, Y. J., & Chen, H. M., (2017). *Chemical Society Reviews, 46*, 337–365.
40. Xu, H., Cheng, D., Cao, D., & Zeng, X. C., (2018). *Nature Catalysis, 1*, 339.
41. Zhang, M., Li, C., Qu, L., Fu, M., Zeng, G., Fan, C., Ma, J., & Zhan, F., (2014). *Applied Surface Science, 300*, 58–65.
42. Sun, J., Fang, Y. H., & Liu, Z. P., (2014). *Physical Chemistry Chemical Physics, 16*, 13733–13740.
43. Guo, D., Shibuya, R., Akiba, C., Saji, S., Kondo, T., & Nakamura, J., (2016). *Science, 351*, 361–365.
44. Jamesh, M. I., & Sun, X., (2018). *Journal of Power Sources, 400*, 31–68.
45. Tahir, M., Pan, L., Idrees, F., Zhang, X., Wang, L., Zou, J. J., & Wang, Z. L., (2017). *Nano Energy, 37*, 136–157.
46. Nørskov, J. K., Rossmeisl, J., Logadottir, A., Lindqvist, L., Kitchin, J. R., Bligaard, T., & Jonsson, H., (2004). *The Journal of Physical Chemistry B, 108*, 17886–17892.
47. Mao, X., Zhang, L., Kour, G., Zhou, S., Wang, S., Yan, C., Zhu, Z., & Du, A., (2019). *ACS Applied Materials and Interfaces*.
48. Li, M., Zhang, L., Xu, Q., Niu, J., & Xia, Z., (2014). *Journal of Catalysis, 314*, 66–72.
49. Guan, J., Duan, Z., Zhang, F., Kelly, S. D., Si, R., Dupuis, M., Huang, Q., et al., (2018). *Nature Catalysis, 1*, 870.
50. Li, X. F., Li, Q. K., Cheng, J., Liu, L., Yan, Q., Wu, Y., Zhang, X. H., et al., (2016). *Journal of the American Chemical Society, 138*, 8706–8709.

51. Riyaz, M., & Goel, N., (2019). *ChemPhysChem*, *20*, 1954–1959.
52. Tian, Y. H., Hu, S., Sheng, X., Duan, Y., Jakowski, J., Sumpter, B. G., & Huang, J., (2018). *The Journal of Physical Chemistry Letters*, *9*, 570–576.
53. Somorjai, G. A., & Li, Y., (2010). *Introduction to Surface Chemistry and Catalysis*. John Wiley & Sons.
54. Bohme, D. K., & Schwarz, H., (2005). *Angewandte Chemie International Edition*, *44*, 2336–2354.
55. Allen, G. C., El-Turki, A., Hallam, K. R., McLaughlin, D., & Stacey, M., (2004). In: *Stone Deterioration in Polluted Urban Environments* (pp. 135–146). CRC Press.
56. Wu, X., Lin, F., Xu, H., & Weng, D., (2010). *Applied Catalysis B: Environmental*, *96*, 101–109.
57. Fajın, J. L., Cordeiro, M. N. D., & Gomes, J. R., (2011). *Chemical Physics Letters*, *503*, 129–133.
58. Thompson, T. L., & Yates, J. T., (2006). *Chemical Reviews*, *106*, 4428–4453.
59. Xie, Y., Dong, F., Heinbuch, S., Rocca, J. J., & Bernstein, E. R., (2010). *Physical Chemistry Chemical Physics*, *12*, 947–959.
60. Kumari, I., Gupta, S., & Goel, N., (2016). *Computational and Theoretical Chemistry*, *1091*, 107–114.
61. Zhao, L., Tan, M., Chen, J., Ding, Q., Lu, X., Chi, Y., Yang, G., et al., (2013). *The Journal of Physical Chemistry A*, *117*, 5161–5170.
62. Wang, Z. C., Yin, S., & Bernstein, E. R., (2013). *The Journal of Physical Chemistry A*, *117*, 2294–2301.
63. Feyel, S., Dobler, J., Schroder, D., Sauer, J., & Schwarz, H., (2006). *Angewandte Chemie*, *118*, 4797–4801.
64. Wu, X. N., Xu, B., Meng, J. H., & He, S. G., (2012). *International Journal of Mass Spectrometry*, *310*, 57–64.
65. Ding, X. L., Wu, X. N., Zhao, Y. X., & He, S. G., (2011). *Accounts of Chemical Research*, *45*, 382–390.
66. Schroeder, D., Fiedler, A., Hrusak, J., & Schwarz, H., (1992). *Journal of the American Chemical Society*, *114*, 1215–1222.
67. Kretzschmar, I., Fiedler, A., Harvey, J. N., Schroder, D., & Schwarz, H., (1997). *The Journal of Physical Chemistry A*, *101*, 6252–6264.
68. Feyel, S., Dobler, J., Schroder, D., Sauer, J., & Schwarz, H., (2006). *Angewandte Chemie International Edition*, *45*, 4681–4685.
69. Fiedler, A., Kretzschmar, I., Schroder, D., & Schwarz, H., (1996). *Journal of the American Chemical Society*, *118*, 9941–9952.
70. Wu, X. N., Ding, X. L., Li, Z. Y., Zhao, Y. X., & He, S. G., (2014). *The Journal of Physical Chemistry C*, *118*, 24062–24071.
71. Zemski, K., Justes, D., & Castleman, A., (2001). *The Journal of Physical Chemistry A*, *105*, 10237–10245.
72. Rozanska, X., & Sauer, J., (2009). *The Journal of Physical Chemistry A*, *113*, 11586–11594.
73. Bell, R. C., & Castleman, A., (2002). *The Journal of Physical Chemistry A*, *106*, 9893–9899.
74. Dong, F., Heinbuch, S., Xie, Y., Bernstein, E. R., Rocca, J. J., Wang, Z. C., Ding, X. L., & He, S. G., (2009). *Journal of the American Chemical Society*, *131*, 1057–1066.

75. Wang, Z. C., Xue, W., Ma, Y. P., Ding, X. L., He, S. G., Dong, F., Heinbuch, S., et al., (2008). *The Journal of Physical Chemistry A, 112*, 5984–5993.
76. Weckhuysen, B. M., & Schoonheydt, R. A., (1999). *Catalysis Today, 51*, 215–221.
77. Weckhuysen, B. M., & Schoonheydt, R. A., (1999). *Catalysis Today, 51*, 223–232.
78. Hogan, J., & Banks, R., (1956). *Ind. Eng. Chem, 48*, 1152.
79. Kimbrough, D. E., Cohen, Y., Winer, A. M., Creelman, L., & Mabuni, C., (1999). *Critical Reviews in Environmental Science and Technology, 29*, 1–46.
80. Catti, M., Sandrone, G., Valerio, G., & Dovesi, R., (1996). *Journal of Physics and Chemistry of Solids, 57*, 1735–1741.
81. Goel, N., (2018). *Chemical Modeling, 14*, 126.
82. Dong, F., Heinbuch, S., Xie, Y., Rocca, J. J., Bernstein, E. R., Wang, Z. C., Deng, K., & He, S. G., (2008). *Journal of the American Chemical Society, 130*, 1932–1943.
83. Weber, M., Weber, M., & Kleine-Boymann, M., (2004). Phenol. *Ullmann's Encyclopedia of Industrial Chemistry*.
84. Wallace, J., (2000). *Kirk-Othmer Encyclopedia of Chemical Technology*. John Wiley & Sons, Inc.
85. Yoshizawa, K., Shiota, Y., & Yamabe, T., (1999). *Journal of the American Chemical Society, 121*, 147–153.
86. Fellah, M. F., Van, S. R. A., & Onal, I., (2009). *The Journal of Physical Chemistry C, 113*, 15307–15313.
87. Altinay, G., & Metz, R. B., (2010). *Journal of the American Society for Mass Spectrometry, 21*, 750–757.
88. Shiota, Y., Suzuki, K., & Yoshizawa, K., (2005). *Organometallics, 24*, 3532–3538.
89. Fellah, M. F., Onal, I., & Van, S. R. A., (2010). *The Journal of Physical Chemistry C, 114*, 12580–12589.
90. Kaur, N., Kumari, I., Gupta, S., & Goel, N., (2016). *The Journal of Physical Chemistry A, 120*, 9588–9597.
91. Xi, Y., Chen, B., Lin, X., Wang, C., & Fu, H., (2016). *Journal of Molecular Modeling, 22*, 79.
92. Xi, Y., Chen, B., Lin, X., Fu, H., & Wang, C., (2016). *Computational and Theoretical Chemistry, 1076*, 65–73.
93. Cherian, M., Rao, M. S., Hirt, A. M., Wachs, I. E., & Deo, G., (2002). *Journal of Catalysis, 211*, 482–495.
94. Wang, S., Murata, K., Hayakawa, T., Hamakawa, S., & Suzuki, F., (2000). *Applied Catalysis A: General, 196*, 1–8.
95. Michorczyk, P., Pietrzyk, P., & Ogonowski, J., (2012). *Microporous and Mesoporous Materials, 161*, 56–66.
96. Weckhuysen, B. M., Wachs, I. E., & Schoonheydt, R. A., (1996). *Chemical Reviews, 96*, 3327–3350.
97. Gaspar, A., Brito, J., & Dieguez, L., (2003). *Journal of Molecular Catalysis A: Chemical, 203*, 251–266.
98. Hakuli, A., Harlin, M., Backman, L., & Krause, A., (1999). *Journal of Catalysis, 184*, 349–356.
99. Michorczyk, P., & Ogonowski, J., (2003). *Applied Catalysis A: General, 251*, 425–433.

100. Ramani, N. C., Sullivan, D. L., Ekerdt, J. G., Jehng, J. M., & Wachs, I. E., (1998). *Journal of Catalysis, 176*, 143–154.

101. Tatibouet, J. M., Wachs, I. E., et al., (1992). *Journal of Catalysis, 136*, 209–221.

102. Kim, D., & Wachs, I. E., (1993). *Journal of Catalysis, 142*, 166–171.

103. Gao, J., Zheng, Y., Tang, Y., Jehng, J. M., Grybos, R., Handzlik, J., Wachs, I. E., & Podkolzin, S. G., (2015). *ACS Catalysis, 5*, 3078–3092.

104. Liotta, L., Venezia, A., Pantaleo, G., Deganello, G., Gruttadauria, M., & Noto, R., (2004). *Catalysis Today, 91*, 231–236.

105. Thu¨ne, P., Linke, R., Van, G. W., De Jong, A., & Niemantsverdriet, J., (2001). *The Journal of Physical Chemistry B, 105*, 3073–3078.

106. Handzlik, J., Grybos, R., & Tielens, F., (2013). *The Journal of Physical Chemistry C, 117*, 8138–8149.

107. Kumari, I., Kaur, N., Gupta, S., & Goel, N., (2019). *Journal of Molecular Modeling, 25*, 17.

CHAPTER 6

Effect of Dye Concentration on Series Resistance of Thionin Dye-Based Organic Diode

PALLAB KUMAR DAS,[1] SWAPAN BHUNIA,[2] SARMISTHA BASU,[3] and N. B. MANIK[4]

[1]*Assistant Professor, Department of Electronics, Behala College, Behala, Kolkata – 700060, West Bengal, India*

[2]*Assistant Professor, Department of Physics, Ramakrishna Residential College (Autonomous), Narendrapur, Kolkata – 700103, West Bengal, India*

[3]*Lecturer, Department of Electronics, Behala College, Behala, Kolkata – 700060, West Bengal, India*

[4]*Condensed Matter Physics Research Center, Department of Physics, Jadavpur University, Kolkata – 700032, India, Tel.: +91 9831209230, E-mail: nb_manik@yahoo.co.in*

ABSTRACT

The series resistance (R_s) plays a significant role in the organic device performance. The typical value of R_s in any organic device is very high which is mainly because of the interface and the trapping of charge carriers at the bulk and also at the barrier potential at the interface. In this work, we have prepared Thionin dye-based organic diode and studied R_s by varying the dye concentration of the diode. To prepare this diode a thin layer of Thionin dye of different concentrations of 2 mg, 4 mg, 6 mg, and 8 mg are sandwiched separately in between two electrodes one of which is ITO coated glass and another is Al. The dark current-voltage (I–V) characteristics of these diodes have been studied and estimated the values of R_s by using Cheung Cheung function. It is observed that the estimated value of R_s for 2 mg is very high and by increasing the dye concentration, the value of R_s is reduced. It is also observed that the value of the ideality factor (η) also reduced by increasing the dye concentration.

6.1   INTRODUCTION

Recently, organic materials are taking an important role in the research field of materials science due to the discovery of conductivity in those materials [1, 2]. Different organic and polymer materials are being widely used to develop different electronic and optoelectronic devices such as a diode, photodiode, solar cell, etc. Devices based on these organic materials have advantages over inorganic devices such as their low cost, preparation techniques, easy tunability of electrical and optical properties [3–6]. In organic devices, one can easily dope to modify its electrical and optical properties. Though there are some limitations due to its low mobility, presence of traps, moisture effects, etc. Due to these limitations, low efficiency is a major problem in an organic device [7, 8]. Basically, in these devices, an active layer of different organic/polymer materials are sandwiched in between two electrodes. The electrical properties are the major key factors to control the device operation and this depends on the junction properties which again depend on the active layer and the electrode materials. Use of different electrodes of different work functions changes the device operation. Moreover, the organic materials are disorder amorphous solid and there is no such regular lattice structure. So due to structural difference and various limitations, the behavior of organic devices are different from inorganic devices [9]. It was already discussed that the current in any organic device is very low and also by increasing the dye concentration, the device efficiency can be improved. Moreover, it was also reported that the dark I–V characteristics of any organic diode deviates from linearity in the logarithm scale which is because of the effect of R_s [10–12]. Values of these parameters like R_s, ideality factor (η) have unusually high value in the organic diode which influences device performance [17, 18]. So it can understand that by reducing R_s the device performance can be improved. In our present work, we prepared an ITO/Thionin/Aldiode, and the dark I–V characteristics are studied of this system by changing the dye concentration. The data has been fitted with Cheung Chung method which is modification of thermionic emission (TE) process by including series resistance and from these fitting series resistance (R_s) and ideality factor (η) have been estimated.

6.2 EXPERIMENTAL DETAILS

6.2.1 MATERIALS USED

Thionin is a strongly metachromatic dye, useful for the staining of acid mucopolysaccharides. It is also a common nuclear stain and can be used for the demonstration of Nissl substance in nerve cells of the CNS. The plant peptide family of thionins normally consists of 45–48 amino acids of which 6–8 are cysteins. These form 3–4 disulfide bridges stabilizing an L-shape with two anti-parallel alpha-helices as the long axis and a small anti-parallel beta-strand as the short axis. There is, apparently, another dye named thionin blue (CI 52025) which is very occasionally confused with thionin, but which cannot be substituted for it. The structure of the dye is shown in Figure 6.1.

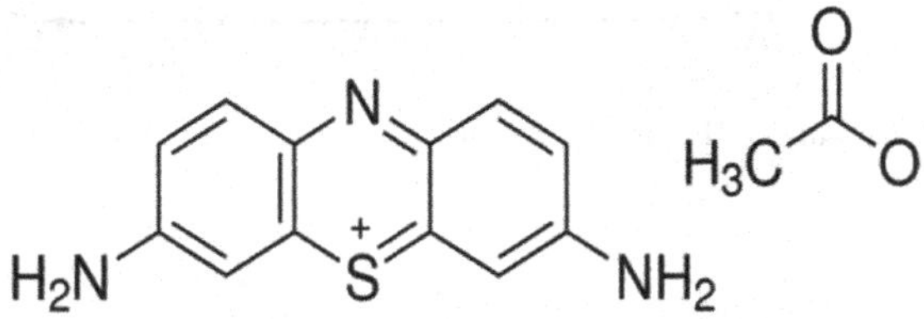

FIGURE 6.1 Structure of thionin dye.

6.2.2 SAMPLE AND CELL PREPARATION

In a cleaned test tube 1 gm of polyvinyl alcohol (PVA) was mixed with 15 cc of distilled water, warmed gently, and stirred to make a transparent viscous solution of PVA. A 2 mg of thionin dye was mixed with this solution. A solid electrolyte was prepared in another cleaned beaker by mixing poly-ethylene oxide (PEO), lithium perchlorate ($LiClO_4$), ethylene carbonate (EC), and propylene carbonate (PC). The mixture of PEO-$LiClO_4$-EC-PC were mixed (30.60%, 3.60%, 46.20%, 19.60% by weight of PVA) stirred and heated for about 5 hrs. This gel-like solid is now mixed with the previously prepared solution of PVA and dye to form the blend. This blend is heated and stirred properly to mix them well. This viscous gel solution is sandwiched between transparent Indium tin oxide (ITO) coated glass plate and Aluminum plate. The first blend of materials spin-coated on ITO and after that Al plate is placed on it to have sandwiched structure. These two plates act as the two contact electrodes. Two electrical leads are then taken out of the two ends of the electrodes. The complete cell is then dried under vacuum for about 6 hrs. Now the cell is ready for characterization. For I–V measurement a Keithley 420 source-measure unit was issued. The structure of the prepared cell is shown in Figures 6.2a and 6.2b.

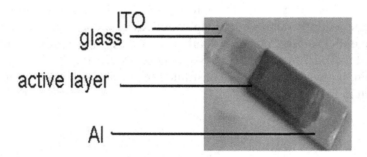

FIGURE 6.2A   Structure of the prepared cell.

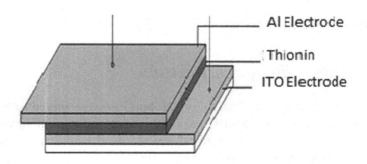

FIGURE 6.2B   Schematic diagram of the prepared ITO/thionin/Al cell.

6.2.3 MEASUREMENTS

The circuit diagram is shown in Figure 6.3. The forward bias corresponds to the positive terminal to ITO with respect to Al.

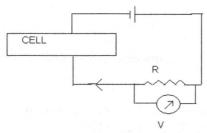

FIGURE 6.3 Circuit diagram for the measurement of the dark current-voltage characteristics.

6.2.4 RESULTS AND DISCUSSION

In order to characterize a diode, the most simple, and useful method is the current-voltage (I–V) characterization. This method is used to extract the main parameters of diode such as ideality factor (n), series resistance, and barrier height. The measured forward bias I–V characteristics of the ITO/thionin/Al diode by varying different dye concentration at room temperature are shown in Figure 6.4.

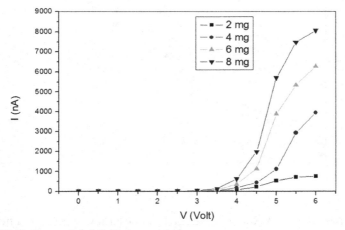

FIGURE 6.4 Forward dark I–V characteristics.

The electrical characterizations were performed at room temperature containing the current-voltage (I–V) measurements which carried out in dark. According to the TE theory [14], the forward bias I–V characteristics of diode contact for qV>3kT can be expressed as:

$$I = I_0 \left[\exp\left(\frac{qV}{\eta kT} \right) - 1 \right] \tag{1}$$

where, I_0 is the reverse saturation current can be described as:

$$I_0 = AA^* \exp(q\Phi_b / KT) \tag{2}$$

Here A is the contact area, A^* is the Richardson constant, T is the absolute temperature, and Φ_b is barrier height.

Ideality factor can be calculated from the I–V characteristic. Figure 6.5 shows the logarithmic values of current versus voltage of this device. The ideality factor η can be obtained from the slope of the IV characteristic by using Eqn. (3).

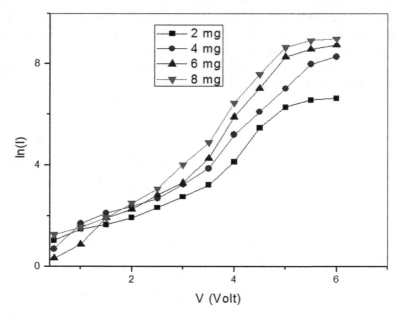

FIGURE 6.5   ln(I) vs V characteristics of ITO/Thionine\Al system for different dye concentrations.

$$\eta= \frac{q}{kT}\frac{dV}{d\ln I} \qquad (3)$$

As we know, the ideality factor (η) measures the conformity of the diode to pure TE. In our ITO/thionine/Al system η is very high. There are various factors which may be responsible for the greater value of (η) for the diode [15, 16]. The extracted values of η for different dye concentration are shown in the table below:

System	Dye Concentration (mg)	Extracted Values of η
ITO/thionin/ Al	2.0	13.07
	4.0	11.96
	6.0	9.90
	8.0	9.20

It is well known that the forward bias I–V characteristic at sufficiently high voltage shows a prominent downward concave nature which indicates the presence of the effect of R_s [17, 18]. If the series resistance effect is low, the non-linear region will be narrow. The values of the R_s were calculated using a method developed by Cheung and Cheung [19, 20]. According to Cheung and Cheung, the forward bias I–V characteristics due to the TE of a diode with the series resistance can be expressed as:

$$I= I_0 \left[\exp\left(\frac{q(V-IRs)}{nkT}\right)-1\right] \qquad (4)$$

where, the IR_s term is the voltage drop across series resistance of the device. The values of the series resistance (R_s) can be extracted from the following equations (5):

$$\frac{dV}{d\ln I} = \frac{nkT}{q}+ IR_s \qquad (5)$$

$$H(I)=V-\left(\frac{nKT}{q}\right)\ln\frac{I}{AA^*T^2} \qquad (6)$$

A plot of dV/d(lnI) vs. I will be linear and gives R_s as the slope and nkT/q as the y-axis intercept from Eqn. (5). Plots of dV/d(lnI) vs. I and H(I) vs. I at room temperature have shown in Figure 6.6(a) to 6.5(h).

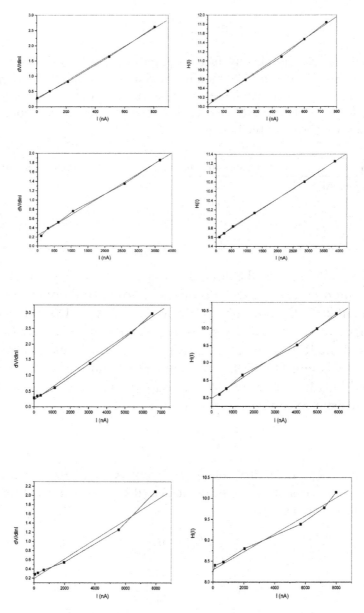

FIGURE 6.6   dV/d(lnI)-Current(I) and H(I)-Current(I) plot for different dye concentration; (a) and (b): for 2 mg dye concentration, (c) and (d): for 4 mg dye concentration, (e) and (f): for 6 mg dye concentration, (g) and (h): for 8 mg dye concentration.

The extracted values of η and R_s by using Cheung method are shown in the table below. It has been observed with increasing the dye concentration the value of R_s is reduced. Moreover, there is a small difference between the values of η obtained from the forward bias $\ln I–V$ plot and from the $dV/d(\ln I)–I$ curves. This may be attributed to the existence of the series resistance and interface states and to the voltage drop across the interfacial layer.

Dye Concentration (mg)	Value of η	Value of R_s (MΩ)
2.0	9.3	2.90
4.0	8.9	0.45
6.0	7.9	0.40
8.0	7.2	0.20

As organic semiconductors are disorder in nature and the molecules are held together by weak Van der Waal's force, these organic materials do not have clearly defined band structure [21]. The energy band is defined by HOMO and LUMO where HOMO is the highest occupied molecular orbital and LUMO stands for lowest unoccupied molecular orbital. Due to this structural disorder and the weak molecular bonding, the organic materials are prone to have electronic traps which introduce additional energy levels inside the energy band between HOMO and LUMO [22, 23].

In ITO/thionin/Al system the interface between the organic dye and the electrodes plays an important role in the device performance. In Figure 6.7, a schematic diagram of the metal-organic dye interface is shown. From Figure 6.7 is seen that when they are connected to fabricate the device depending on the work function charge carriers will diffuse until the Fermi levels are aligned. Due to this transfer of charge, there is a band bending effect that occurs at the metal-semiconductor interface.

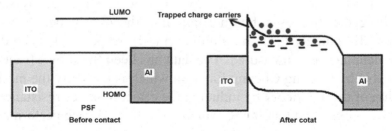

FIGURE 6.7 Band structure of ITO/Thionin/Al organic diode.

The LUMO level shifts down near the ITO/thionin interface due to the transfer of holes from ITO to thionin. The LUMO level also goes up near the thionin/Al interface due to the transfer of electrons from Al to thionin. The band bending effect for the migration of charge carriers produce a potential barrier near interfaces. Due to this barrier, there is no flow of charge carriers at equilibrium. In this case, a forward bias is applied by connecting the positive terminal of the source to ITO and the negative terminal to Al. When the applied voltage is greater than the potential barrier of the interface the carriers are transported. But as organic semiconductors are disorder in nature and very much prone to traps, a large part of the charge carriers injected from the electrode are being trapped in these trap energy states at the interface which is shown in Figure 6.6. So the carriers will not come out from the electrode and the current, as well as the mobility of the charge carriers, becomes very low. At higher voltage, these trapped carriers in between HOMO and LUMO may be released or recombined with the opposite charge carriers which increase the current for the device. But due to the disorder of the organic material, presence of traps, recombination of carriers at the metal-semiconductor interface, the current flow through the device is very low. So the values of electronic parameters η, R_s of this organic device become unusually high values. Also, it is expected that the charge trapping in the metal-organic dye interface is responsible for this high value of R_s By increasing the dye concentration, the no of charge carriers are increased which in turn increase the device performance.

6.3   CONCLUSIONS

In this work, we have analyzed the I–V characteristics of an ITO/thionin/Al organic diode. The diode has been fabricated by sandwiching the viscous solution containing the dye in between two electrodes namely ITO and aluminum (Al), respectively. The dark I–V data has been measured. The data has been fitted by TE theory and also by Cheung Cheung function which is basically the modification of the TE theory by inducing the effect of series resistance. It has been observed that the R_s which control the device performance significantly can be reduced by increasing the dye concentration. The extracted values of R_s are about 2.9 MΩ, 0.45 MΩ, 0.40 MΩ, and 0.20 MΩ, respectively, for the dye concentration of 2 mg, 4 mg, 6 mg, and 8 mg. Moreover, From the TE theory the ideality factor is calculated from the linear portion of the lnI Vs V plot and by using Cheung-Cheung function η of the device has been estimated from the Y-axis of the dV/d(lnI) vs. I plot. It is found that in both the cases η is quite high but by considering series resistance effect this value decreases and with reduction of series resistance the ideality factor (η) also been reduced.

KEYWORDS

- **electrochemical cell**
- **ethylene carbonate**
- **ideality factor (η)**
- **polyethylene oxide**
- **polyvinyl alcohol**
- **series resistance (R_s)**
- **thermionic emission**

REFERENCES

1. Donald, M. C., McCarthy, D. K., & Genz, R. H., (1964). The effect of structure on the electrical conductivity of organic compounds polyazophenylenes. *The Journal of Physical Chemistry, 68*(9), 2661–2665.
2. Stenger-Smith, J. D., (1998). Intrinsically electrically conducting polymers. Synthesis, characterization, and their applications. *Progress in Polymer Science, 23*(1), 57–59.
3. Dodabalapur, A., (1997). Organic light emitting diodes. *Solid State Communications, 102*(2), 259–267.
4. Saleem, M., Sayyad, M. H., Ahmad, Z., & Karimove, K. S., (*2007*). Fabrication and investigation of the charge/ discharge characteristics of zinc/PVA-KOH-H$_2$O-Iodine/ carbon cell. *Optoelectronics and Advanced Materials-Rapid Communication, 1(9), 477–479.*

5. Ajanta, H., Subhasis, M., & Manik, N. B., (2009). Electrical and photovoltaic characterizations of methyl red dye doped solid-state photoelectrochemical cell. *Ionics, 15*(1), 79–83.

6. Ishii, M., & Taga, Y., (2002). Influence of temperature and drive current on degradation mechanisms in organic light-emitting diodes. *Applied Physics Letters, 80*(18), 3430–3432.

7. Chakraborty, S., & Manik, N. B., (2014). Effect of COOH-functionalized SWCNT addition on the electrical and photovoltaic characteristics of malachite green dye based photovoltaic cells. *Journal of Semiconductors, 35*(12), 124004(1–6).

8. Hagberg, D. P., Marinado, T., Karlsson, K. M., Nonomura, K., Qin, P., et al., (2007). Tuning the HOMO and LUMO energy levels of organic chromophores for dye sensitized solar cells. *Journal of Organic Chemistry, 72*(25), 9550–9556.

9. Li, G., Zhu, R., & Yang, Y., (2012). Polymer solar cells. *Nature Photonics, 6*(3), 153–161.

10. Khasan, K. S., Mahroof-Tahir, M., Saleem, M., et al., (2011). I–V characteristics of vanadium-flavonoid complexes based Schottky diodes. *Physica. B, 406*(15), 3011–3017.

11. Aydogan, S., Saglam, M., & Turut, A., (2008). Some electrical properties of polyaniline/p-Si/Al structure at 300 K and 77 K temperatures. *Microelectronic Engineering, 85*(2), 278–283.

12. Gullu, O., Aydogan, S., & Turut, A., (2008). Fabrication and electrical characteristics of Schottky diode based on organic material. *Microelectronic Engineering, 85*(7), 1647–1651.

13. Saha, I., Hossain, M., & Suresh, K. G., (2010). Sequence-Selective binding of phenazinium dyes phenosafranin and safranin O to guanine-cytosine deoxyribo polynucleotides: Spectroscopic and thermodynamic studies. *The Journal of Physical Chemistry B, 114*(46), 15278–15287.

14. Rhoderick, E. H., (1978). *Metal-Semiconductor Contacts.* Oxford University Press.

15. Aydoğan, S., Incekara, Ü., Deniz, A. R., & Türüt, A., (2010). Extraction of electronic parameters of Schottky diode based on an organic indigotindisulfonate sodium (IS). *Solid State Communications, 150*(33), 1592–1596.

16. Ahmad, Z., & Sayyad, M. H., (2009). Extraction of electronic parameters of Schottky diode based on an organic semiconductor methyl-red. *Physica. E, 41*(4), 631–634.

17. Ahmad, Z., & Sayyad, M. H., (2009). Electrical characteristics of a high rectification ratio organic Schottky diode based on methyl red. *Optoelectronics and Advanced Materials Rapid Communications, 3*(5), 509–512.

18. Mutabar, S., Sayyad, M. H., & Karimov, K. S., (2011). Electrical characterization of the organic semiconductor Ag/CuPc/Au Schottky diode. *Journal of Semiconductors, 32*(4), 044001(1–5).

19. Cheung, S. K., & Cheung, N. W., (1986). Extraction of Schottky diode parameters from forward current-voltage characteristics. *Applied Physics Letters, 49*(2), 85–87.

20. Paoli, L., & Barnes, P. A., (1976). Saturation of the junction voltage in stripe-geometry (AlGa)As double-heterostructure junction lasers. *Applied Physics Letters, 28*(12), 714–716.

21. Jie, Y., & Jun, S., (1999). Effects of discrete trap levels on organic light emitting diodes. *J. Appl. Phys., 85(5)*, 2699–2705.

22. Islam, M. R., Saha, S., Manik, N. B., & Basu, A. N., (2012). Transient current study in safranine-T dye based organic photo-electrochemical cell using exponentially distributed trap assisted charge transport model. *Indian J. Phys., 86(12)*, 1101–1106.

23. Walker, A. B., Peter, L. M., Martínez, D., & Lobato, K., (2007). Transient photocurrents in dye-sensitized nanocrystalline solar cells. *CHIMIA International Journal for Chemistry, 61(12)*, 792–795.

CHAPTER 7

On Gower's Inverse Matrix

J. LÓPEZ-BONILLA, R. LÓPEZ-VÁZQUEZ, and S. VIDAL-BELTRÁN

Superior School of Electrical and Mechanical Engineering, National Polytechnic Institute, Building 4, Lindavista CP 07738, Mexico City, Mexico, E-mail: jlopezb@ipn.mx (J. López-Bonilla)

ABSTRACT

We show that Faddeev-Sominsky's process allows construct a natural inverse for any square matrix, which is an alternative to the inverse obtained by Gower.

7.1   INTRODUCTION

For an arbitrary matrix $\mathbf{A}_{nxn} = \left(A^i{}_j \right)$ its characteristic equation [1–3]:

$$\lambda^n + a_1 \lambda^{n-1} + \cdots + a_{n-1}\lambda + a_n = 0, \tag{1}$$

Can be obtained, through several procedures [1, 4–7], directly from the condition $\det \left(A^i{}_j - \lambda \delta^i_j \right) = 0$. The approach of Leverrier-Takeno [4, 8–12] is a simple and interesting technique to construct (1) based in the traces of the powers $\mathbf{A}^r$, $r = 1, \ldots, n$. On the other hand, it is well known that an arbitrary matrix satisfies (1), which is the Cayley-Hamilton-Frobenius identity [1–3]:

$$\mathbf{A}^n + a_1 \mathbf{A}^{n-1} + \cdots + a_{n-1}\mathbf{A} + a_n\mathbf{I} = 0. \tag{2}$$

If $\mathbf{A}$ is non-singular (that is, $\det \mathbf{A} \neq 0$), then from (2) we obtain its inverse matrix:

$$\mathbf{A}^{-1} = -\frac{1}{a_n}\left(\mathbf{A}^{n-1} + a_1 \mathbf{A}^{n-2} + \cdots + a_{n-1}\mathbf{I} \right), \tag{3}$$

where, $a_n \neq 0$ because $a_n = (-1)^n \det \mathbf{A}$.

Faddeev-Sominsky [13–15] proposed an algorithm to determine $\mathbf{A}^{-1}$ in terms of $\mathbf{A}^r$ and their traces, which is equivalent [16] to the Cayley-Hamilton-Frobenius theorem (2) plus the Leverrier-Takeno's method to construct the characteristic polynomial of a matrix $\mathbf{A}$, to see Section 7.2. The condition $\det \mathbf{A} \neq 0$ means $p \equiv rank \ \mathbf{A} = n$, then it is immediate the quest of an inverse of $\mathbf{A}$ when $1 \leq p \leq (n{-}1)$, in fact, in Section 7.3 we exhibit a pseudoinverse $\mathbf{A}^-$ with the properties [17, 18]:

$$\mathbf{A}\,\mathbf{A}^-\mathbf{A} = \mathbf{A}, \tag{4}$$

$$\mathbf{A}^-\mathbf{A}\,\mathbf{A}^- = \mathbf{A}^-, \tag{5}$$

which we consider more simple and natural than the Gower's inverse [19].

7.2   LEVERRIER-TAKENO AND FADDEEV-SOMINSKY TECHNIQUES

If we define the quantities:

$$a_0 = 1, \qquad s_k = tr \ \mathbf{A}^k, \quad k = 1, 2, \ldots, n \tag{6}$$

Then the process of Leverrier-Takeno [4, 8–12] implies (1) where in the a_i are determined with the recurrence relation:

$$r\,a_r + s_1 a_{r-1} + s_2 a_{r-2} + \ldots + s_{r-1} a_1 + s_r = 0, \quad r = 1,2,\ldots,n \tag{7}$$

Therefore:

$$a_1 = -s_1 \ , \qquad 2!a_2 = \left(s_1\right)^2 - s_2 \ , \qquad 3!a_3 = -\left(s_1\right)^3 + 3 s_1 s_2 - 2 s_3 \ ,$$

$$4!a_4 = \left(s_1\right)^4 - 6\left(s_1\right)^2 s_2 + 8 s_1 s_3 + 3\left(s_2\right)^2 - 6 s_4 \ , \quad etc. \tag{8}$$

In particular, $\det \mathbf{A} = (-1)^n a_n$, that is, the determinant of any matrix only depends on the traces s_r, which means that $\mathbf{A}$ and its transpose have the same determinant. In Refs. [20, 21] we find the general expression:

$$a_k = \frac{(-1)^k}{k!} \begin{vmatrix} s_1 & k-1 & 0 & \cdots & 0 \\ s_2 & s_1 & k-2 & \cdots & 0 \\ \vdots & \vdots & \vdots & \ddots & \vdots \\ s_{k-1} & s_{k-2} & \cdots & \cdots & 1 \\ s_k & s_{k-1} & \cdots & \cdots & s_1 \end{vmatrix}, \qquad k=1,\ldots,n . \qquad (9)$$

The Faddeev-Sominsky's procedure [13–16, 19, 22] to obtain $\mathbf{A}^{-1}$ is a sequence of algebraic computations on the powers $\mathbf{A}^r$ and their traces, in fact, this algorithm is given via the instructions:

$$\mathbf{B}_1 = \mathbf{A}, \qquad q_1 = \operatorname{tr} \mathbf{B}_1, \qquad \mathbf{C}_1 = \mathbf{B}_1 - q_1 \mathbf{I},$$

$$\mathbf{B}_2 = \mathbf{C}_1 \mathbf{A}, \qquad q_2 = \frac{1}{2} \operatorname{tr} \mathbf{B}_2, \qquad \mathbf{C}_2 = \mathbf{B}_2 - q_2 \mathbf{I}, \qquad (10)$$

$$\mathbf{B}_{n-1} = \mathbf{C}_{n-2} \mathbf{A}, \qquad q_{n-1} = \frac{1}{n-1} \operatorname{tr} \mathbf{B}_{n-1}, \qquad \mathbf{C}_{n-1} = \mathbf{B}_{n-1} - q_{n-1} \mathbf{I},$$

$$\mathbf{B}_n = \mathbf{C}_{n-1} \mathbf{A}, \qquad q_n = \frac{1}{n} \operatorname{tr} \mathbf{B}_n, \qquad \mathbf{C}_n = \mathbf{B}_n - q_n \mathbf{I} = 0,$$

Then:

$$\mathbf{A}^{-1} = \frac{1}{q_n} \mathbf{C}_{n-1}. \qquad (11)$$

For example, if we apply (10) for $n = 4$, then it is easy to see that the corresponding qr imply (6) with $q_j = -a_j$, and besides (11) reproduces (3). By mathematical induction one can prove that (10) and (11) are equivalent to (3), (4) and (5), showing [16] thus that the Faddeev-Sominsky's technique has its origin in the Leverrier-Takeno method plus the Cayley-Hamilton-Frobenius theorem.

From (10) we can see that [22]:

$$\mathbf{C}_k = \mathbf{A}^k + a_1 \mathbf{A}^{k-1} + a_2 \mathbf{A}^{k-2} + \ldots + a_{k-1} \mathbf{A} + a_k \mathbf{I}, \quad k = 1, 2, \ldots, n-1, \qquad (12)$$

And for $k = n-1$:

$$\mathbf{C}_{n-1} = \mathbf{A}^{n-1} + a_1 \mathbf{A}^{n-2} + a_2 \mathbf{A}^{n-3} + \ldots + a_{n-2} \mathbf{A} + a_{n-1} \mathbf{I} = -a_n \mathbf{A}^{-1},$$

In harmony with (11) because $a_n = -q_n$. The property $C_n = 0$ is equivalent to (2). If A is singular, the process (10) gives the adjoint matrix of A [2, 3], in fact, $Adj\ A = (-1)^{n+1} C_{n-1}$.

If the roots of (1) have distinct values, then the Faddeev-Sominsky's algorithm allows obtain the corresponding eigenvectors of A [6]:

$$A\vec{u}_k = \lambda_k \vec{u}_k , \qquad k = 1, 2, \ldots, n, \qquad (13)$$

Because for a given value of k, each column of:

$$Q_k \equiv \lambda_k^{n-1} I + \lambda_k^{n-2} C_1 + \cdots + C_{n-1}, \qquad (14)$$

Satisfies (13), and therefore all columns of Q_k are proportional to each other, that is, $rank\ Q_k = 1$ [19, 23].

7.3   AN ALTERNATIVE TO GOWER'S PSEUDOINVERSE

Here we consider the situation $1 \le p \le (n-1)$ with the condition $a_p \ne 0$, thus the multiplicity of the eigenvalue zero is $(n-p)$ and from (1) we deduce that [19]:

$$a_j = 0, \qquad j = p+1, \ldots, n, \qquad (15)$$

$$\lambda_k^p + a_1 \lambda_k^{p-1} + \cdots + a_{p-1} \lambda_k + a_p = 0, \qquad k = 1, \ldots, p, \qquad (16)$$

Which allows to show the property:

$$A^{p+1} + a_1 A^p + \cdots + a_{p-1} A^2 + a_p A = 0, \qquad (17)$$

If A is non-defective, that is, has n independent eigenvectors. From (10) and (12) we observe that (17) means:

$$C_p A = B_{p+1} = 0, \qquad (18)$$

That is:

$$B_j = C_j = 0, \qquad j = p+1, \ldots, n. \qquad (19)$$

Thus, from (18) we have that:

$$0 = \mathbf{A}\mathbf{C}_p = \mathbf{A}\left(\mathbf{B}_p + a_p\mathbf{I}\right) = \mathbf{A}\mathbf{C}_{p-1}\mathbf{A} + a_p\mathbf{A}$$

$$\therefore \quad \mathbf{A}\left(-\frac{1}{a_p}\mathbf{C}_{p-1}\right)\mathbf{A} = \mathbf{A},$$

Whose comparison with (4) gives the inverse matrix:

$$\mathbf{A}^- = -\frac{1}{a_p}\mathbf{C}_{p-1}, \tag{20}$$

Therefore:

$$\mathbf{A}\mathbf{A}^- = \mathbf{A}^-\mathbf{A} = -\frac{1}{a_p}\mathbf{B}_p. \tag{21}$$

Hence from (4), (12) and (20):

$$\mathbf{A}^-\mathbf{A}\mathbf{A}^- = -\frac{1}{a_p}\,\mathbf{A}^-\mathbf{A}\left(\mathbf{A}^{p-1} + a_1\mathbf{A}^{p-2} + \cdots + a_{p-2}\mathbf{A} + a_{p-1}\mathbf{I}\right) = -\frac{1}{a_p}\mathbf{C}_{p-1} = \mathbf{A}^-,$$

Because $\mathbf{A}^-\mathbf{A}\,\mathbf{A}^k = \mathbf{A}^k$, then (20) verifies (5).
We must note that Gower employs the inverse:

$$\mathbf{A}_G^- = -\frac{1}{a_p}\left(\mathbf{C}_{p-2} - \frac{a_{p-1}}{a_p}\mathbf{C}_{p-1}\right)\mathbf{A} = -\frac{1}{a_p}\left(\mathbf{B}_{p-1} - \frac{a_{p-1}}{a_p}\mathbf{B}_p\right), \tag{22}$$

With the properties (4) and (5), but is immediate to see that:

$$\mathbf{A}_G^-\mathbf{A} = \mathbf{A}^-\mathbf{A}. \tag{23}$$

We consider that our inverse (20) is more simple and natural than (22) because it was implied by the Faddeev-Sominsky's algorithm [13–16, 19, 22, 24].

7.4   CONCLUSIONS

We emphasize that (20) was obtained under the constraints $1 \le p \le (n-1)$, $a_p \ne 0$ and $\mathbf{A}$ non-defective. The Leverrier-Faddeev's technique, according to Householder [24, 25], was rediscovered and improved by Souriau [26] and Frame [27]. It is interesting to mention that the method (10) was successfully applied [28–30] in general relativity to study the embedding of spacetimes into pseudo-Euclidean spaces.

KEYWORDS

- adjoint matrix
- characteristic equation
- eigenvalue problem
- Faddeev-Sominsky's method
- Gower's pseudoinverse matrix
- Leverrier-Takeno's algorithm

REFERENCES

1. Lanczos, C., (1988). *Applied Analysis*. Dover, New York.
2. Hogben, L., (2006). *Handbook of Linear Algebra*. Chapman & Hall / CRC Press, London.
3. Hazra, A. K., (2006). *Matrix: Algebra, Calculus, and Generalized Inverse*. Cambridge Int. Sci. Pub.
4. Wayland, H., (1945). Expansion of determinantal equations into polynomial form. *Quart. Appl. Math., 2*, 277–306.
5. Householder, A. S., & Bauer, F. L., (1959). On certain methods for expanding the characteristic polynomial. *Numerische Math., 1*, 29–37.
6. Wilkinson, J. H., (1965). *The Algebraic Eigenvalue Problem*. Clarendon Press, Oxford.
7. Lovelock, D., & Rund, H., (1975). *Tensors, Differential Forms, and Variational Principles*. John Wiley and Sons, New York.
8. Leverrier, U. J. J., (1840). On the secular variations of elliptical elements of the seven principal planets, *J. de Math. Pures Appl. Série. 1, 5*, 220–254.
9. Krylov, A. N., (1931). On the numerical solution of the equation, that in technical problems, determines the small oscillation frequencies of material systems. *Bull. De l'Acad. Sci. URSS, 7*(4), 491–539.
10. Takeno, H., (1954). A theorem concerning the characteristic equation of the matrix of a tensor of the second order. *Tensor NS, 3*, 119–122.
11. Wilson, E. B., Decius, J. C., & Cross, P. C., (1980). *Molecular Vibrations* (pp. 216–217). Dover, New York.
12. Guerrero-Moreno, I., López-Bonilla, J., & Rivera-Rebolledo, J., (2011). Leverrier-Takeno coefficients for the characteristic polynomial of a matrix. *J. Inst. Eng. (Nepal), 8*(1/2), 255–258.

13. Faddeev, D. K., & Sominsky, I. S., (1949). *Collection of Problems on Higher Algebra.* Moscow.

14. Faddeeva, V. N., (1959). *Computational Methods of Linear Algebra.* Dover, New York, Chap. 3.

15. Faddeev, D. K., (1963). *Methods in Linear Algebra.* W. H. Freeman, San Francisco, USA.

16. Caltenco, J. H., López-Bonilla, J., & Peña-Rivero, R., (2007). Characteristic polynomial of A and Faddeev's method for A^{-1}. *Educatia. Matematica, 3*(1/2), 107–112.

17. Zuhair, N. M., (1976). *Generalized Inverses and Applications.* Academic Press, New York.

18. Ben-Israel, A., & Greville, T. N. E., (2003). *Generalized Inverses: Theory and Applications.* Springer-Verlag, New York.

19. Gower, J. C., (1980). A modified Leverrier-Faddeev algorithm for matrices with multiple eigenvalues. *Linear Algebra and its Applications, 31*(1), 61–70.

20. Brown, L. S., (1994). *Quantum Field Theory* Cambridge University Press.

21. Curtright, T. L., & Fairlie, D. B., (2012). *A Galileon Primer.* arXiv: 1212.6972v1 [hep-th].

22. Hanzon, B., & Peeters, R., (1999–2000). *Computer Algebra in Systems Theory.* Dutch Institute of Systems and Control, Course Program.

23. Cruz-Santiago, R., López-Bonilla, J., & Vidal-Beltrán, S., (2018). *On Eigenvectors Associated to a Multiple Eigenvalue* (Vol. 100, pp. 248–253). World Scientific News.

24. Helmberg, G., Wagner, P., & Veltkamp, G., (1993). On Faddeev-Leverrier's method for the computation of the characteristic polynomial of a matrix and of eigenvectors. *Linear Algebra and its Applications, 185,* 219–233.

25. Householder, A. S., (1964). *The Theory of Matrices in Numerical Analysis.* Blaisdell, New York.

26. Souriau, J. M., (1948). A method for the special decomposition and the inversion of matrices,*C. R. Acad. Sci. Paris, 227,* 1010–1011.

27. Frame, J. S., (1949). A simple recursion formula for inverting a matrix. *Bull. Amer. Math. Soc., 55,* 1045.

28. López-Bonilla, J., & Núñez-Yépez, H., (1996). An identity for space times embedded into E_5. *Pramana J. Phys., 46*(3), 219–221.

29. López-Bonilla, J., Morales, J., & Ovando, G., (2000). An identity for R_4 embedded into E_5. *Indian J. Math., 42*(3), 309–312.

30. López-Bonilla, J., Morales, J., Ovando, G., & Ramírez, E., (2006). Leverrier-Faddeev's algorithm applied to space times of class one. *Proc. Pakistan Acad. Sci., 43*(1), 47–50.

CHAPTER 8

Applications of Noether's Theorem

J. YALJÁ MONTIEL-PÉREZ,[1] J. LÓPEZ-BONILLA,[2] R. LÓPEZ-VÁZQUEZ,[2] and S. VIDAL-BELTRÁN[2]

[1]*Computer Research Center, National Polytechnic Institute, CP 07738, Mexico City, Mexico*

[2]*Superior School of Electrical and Mechanical Engineering, National Polytechnic Institute, Building 4, Lindavista CP 07738, Mexico City, Mexico, E-mail: jlopezb@ipn.mx (J. López-Bonilla)*

ABSTRACT

If the action $S = \int_{t_1}^{t_2} L(q, \dot{q}t)dt$ is invariant under the infinitesimal transformation $\tilde{t} = t + \varepsilon \tau(q,t)$, $\tilde{q}_r = q_r + \varepsilon \xi_r(q,t)$, $r = 1,\ldots,n$, with $\varepsilon = $ constant $\ll 1$, then the Noether's theorem permits to construct the corresponding conserved quantity. The Lanczos approach employs to $\varepsilon = q_{n+1}$ as a new degree of freedom, thus the Euler-Lagrange equation for this new variable gives the Noether's constant of motion. Torres del Castillo and Rubalcava-García showed that each variational symmetry implies the existence of an ignorable coordinate; here we apply the Lanczos approach to the Noether's theorem to motivate the principal relations of these authors. The Maxwell equations without sources are invariant under duality rotations, then we show that this invariance implies, via the Noether's theorem, the continuity equation for the electromagnetic energy. Besides, we demonstrate that if we know one solution of $p(x)y'' + q(x)y' + r(x)y = 0$, then this Lanczos technique allows obtain the other solution of this homogeneous linear differential equation.

8.1   INTRODUCTION

In the functional (the concept of action was proposed by Leibnitz [1]) $S = \int_{t_1}^{t_2} L(q,\dot{q},t)\,dt$ we apply the infinitesimal transformation ($\varepsilon = $ constant $\ll 1$):

$$\tilde{t} = t + \varepsilon\tau(q,t), \qquad \tilde{q}_r = q_r + \varepsilon\xi_r(q,t), \quad r = 1,\ldots,n \tag{1}$$

That is:

$$\tilde{S} = \int_{\tilde{t}_1}^{\tilde{t}_2} L(\tilde{q}, \frac{d\tilde{q}}{d\tilde{t}}, \tilde{t})\,d\tilde{t} \tag{2}$$

Then we say that the action is invariant if:

$$\tilde{S} = S + \varepsilon \int_{t_1}^{t_2} \frac{d}{dt} Q(q,t)\,dt, \tag{3}$$

Hence the Euler-Lagrange equations (Lagrangian expressions [2, 3]) corresponding to the variational principle $\delta S = 0$:

$$E_r \equiv \frac{d}{dt}\left(\frac{\partial L}{\partial \dot{q}_r}\right) - \frac{\partial L}{\partial q_r} = 0, \quad r = 1,\ldots,n \tag{4}$$

Remain intact. Noether [2] studied the case $Q = 0$, and she suggested [3, 4] to Bessel-Hagen [5] the analysis of (4) with $Q \neq 0$ [6].

Therefore, we have a symmetry up to divergence and Noether [2, 5–9] proved the existence of the Rund-Trautman identity [7, 8, 10, 11]:

$$\frac{\partial L}{\partial q_r}\xi_r + \frac{\partial L}{\partial \dot{q}_r}\dot{\xi}_r + \frac{\partial L}{\partial t}\tau - \left(\frac{\partial L}{\partial \dot{q}_r}\dot{q}_r - L\right)\dot{\tau} - \frac{dQ}{dt} = 0, \tag{5}$$

Which can be written in the form:

$$\frac{d}{dt}\left(\frac{\partial L}{\partial \dot{q}_r}\xi_r - H\tau - Q\right) = (\xi_r - \dot{q}_r\tau)E_r, \qquad H = \frac{\partial L}{\partial \dot{q}_c}\dot{q}_c - L. \tag{6}$$

In (5) and (6) we use the convention of Dedekind [12, 13]-Einstein because we sum over repeated indices. The Rund-Trautman identity offers a more efficient test of invariance [8]. If in (6) we employ the Euler-Lagrange equations (4) we deduce the constant of motion associated to (1):

$$\varphi(q,\dot{q},t) \equiv \xi_r - H\tau - Q = Constants, \tag{7}$$

Hence, we have a connection between symmetries and conservation laws [3, 7, 8, 10, 14–16]. We remember the following words of Havas [3, 4]: 'The relation between symmetries and conserved quantities could fail for physical systems whose equations could not be written in Hamiltonian form.' On the importance of the symmetries, we have the comment of Weinberg [17]: 'Wigner realized, earlier than most physicists, the outstanding of thinking about symmetries as objects of interest in themselves, quite apart from the dynamical theory. The symmetries of Nature are the deepest things we know about it.'

In Section 8.2 we exhibit the Lanczos technique [18–24] to obtain the Noether's conserved quantity (7) as the Euler-Lagrange equation for the parameter $\varepsilon(t)$. The Section 8.3 contains a motivation for the result of Torres del Castillo and Rubalcava-García [25], namely, that a variational symmetry implies the existence of an ignorable coordinate. In Section 8.4, we show that the continuity equation for the electromagnetic energy can be deduced from the invariance of Maxwell equations under duality rotations [21]. Section 8.5 has an application of the Noether's theorem to an arbitrary homogeneous linear differential equation of second order [26].

8.2   LANCZOS APPROACH TO CONSERVATION LAWS

Lanczos [18, 19] applies the infinitesimal transformation (1) (with $\varepsilon =$ constant to the action (2) and uses expansion of Taylor up to first order in ε, thus:

$$\tilde{S} = S + \varepsilon \int_{t_1}^{t_2} (\frac{\partial L}{\partial q_r}\xi_r + \frac{\partial L}{\partial \dot{q}_r}\dot{\xi}_r + \frac{\partial L}{\partial t}\tau - H\dot{\tau})dt, \tag{8}$$

Hence this integrand is equal to $\frac{dQ}{dt}$, in harmony with the Rund-Trautman identity (5).

Now Lanczos proposes to employ (1) into (2) but considering that ε is a function, therefore up to 1st order in ε:

$$\tilde{S} = \int_{t_1}^{t_2} [L + \varepsilon \frac{dQ}{dt} + \dot{\varepsilon}\left(\frac{\partial L}{\partial \dot{q}_r}\xi_r - H\tau \right)]dt = \int_{t_1}^{t_2} \overline{L}\, dt, \tag{9}$$

And he accepts that ε is a new degree of freedom with its corresponding Euler-Lagrange equation:

$$\frac{d}{dt}\left(\frac{\partial \bar{L}}{\partial \dot{\varepsilon}}\right) - \frac{\partial \bar{L}}{\partial \varepsilon} = 0.$$ (10)

It is clear that:

$$\frac{\partial \bar{L}}{\partial \dot{\varepsilon}} = \frac{\partial L}{\partial \dot{q}_r}\xi_r - H\tau, \qquad \frac{\partial \bar{L}}{\partial \varepsilon} = \frac{dQ}{dt}$$

Therefore (10) implies (7). In other words, if the parameter of the symmetry is considered as an additional degree of freedom of the variational principle, then its Euler-Lagrange equation gives the Noether's constant of motion. We comment that Neuenschwander [27] obtains the conserved quantity (7) for the case $Q = 0$. If in (8) we use τ and ξ_r as new degrees of freedom, then the corresponding Euler-Lagrange equations imply the equations of motion (4) and the known relation $\frac{dH}{dt} = -\frac{\partial L}{\partial t}$.

The works [20, 22] have applications of this Lanczos technique to some singular Lagrangians employed in [9, 28–30]. In [18, 21] and [31] the Lanczos method is applied to electromagnetic and gravitational fields, respectively.

8.3   IGNORABLE VARIABLES AND CONSTANTS OF MOTION

Here we look a finite transformation of coordinates:

$$t, q_r \rightarrow t', q_r', \quad r = 1,\ldots,n$$ (11)

Such that in the new Lagrangian one coordinate, we say q_1', participates as ignorable variable and its conjugate momentum leads to the constant (7); Torres del Castillo and Rubalcava-García [25] show that (11) can be obtained from the equations:

$$\frac{\partial t}{\partial q_1'} = \tau, \qquad \frac{\partial q_r}{\partial q_1'} = \xi_r, \quad r = 1,\ldots,n$$ (12)

The Lanczos variational technique [18, 19] allows deduce the Noether's conserved quantity (7) as the Euler-Lagrange equation for the parameter

ε if it is considered as a new degree of freedom. Here we employ this Lanczos approach to motivate the relations (12).

Lanczos applies (1) into (2) but considering that ε is a function, therefore up to 1th order in ε:

$$\tilde{S} = \int_{t_1}^{t_2} [L + \dot{\varepsilon}\,\varphi(q,\dot{q},t) + \frac{d}{dt}(\varepsilon Q)]dt = \int_{t_1}^{t_2} \left[\bar{L} + \frac{d}{dt}(\varepsilon Q)\right]dt, \qquad (13)$$

Where we use (5) and (7); thus we can see that $\varepsilon(t)$ is ignorable into $\bar{L}$ and its corresponding Euler-Lagrange equation $\frac{d}{dt}\left(\frac{\partial\bar{L}}{\partial\dot{\varepsilon}}\right) - \frac{\partial\bar{L}}{\partial\varepsilon} = 0$ implies that φ is a constant.

On the other hand, from (1) we have that $t = \tilde{t} - \varepsilon\tau(q,t)$ and $q_r = \tilde{q}_r - \varepsilon\xi_r(q,t)$, then:

$$\frac{\partial t}{\partial\varepsilon} = -\tau\,, \qquad \frac{\partial q_r}{\partial\varepsilon} = -\xi_r\,, \qquad r = 1,\ldots,n \qquad (14)$$

But ε is ignorable and its momentum leads to (7), hence it is natural the identification $\varepsilon = -q_1'$, thus (14) imply the expressions (12) obtained by Torres del Castillo and Rubalcava-García [25], and in their paper, we find several examples on the construction of ignorable variables associated to variational symmetries [32]. Let's remember [18] that the fundamental importance of the ignorable variables for the integration of the Lagrangian equations was first recognized by Routh [33] and Helmholtz [34].

8.4   MAXWELL EQUATIONS AND DUALITY ROTATIONS

The Maxwell equations in the absence of sources are given by [18] $\vec{\nabla}\cdot\vec{E} = 0$, $\vec{\nabla}\cdot\vec{B} = 0$ and:

$$\vec{\nabla}\times\vec{E} + \frac{\partial\vec{B}}{\partial t} = \vec{0}, \qquad \vec{\nabla}\times\vec{B} - \frac{1}{c^2}\frac{\partial\vec{E}}{\partial t} = \vec{0}, \qquad (15)$$

Where $\vec{B}$ and $\vec{E}$ are the magnetic and electric fields, respectively; $c = 1/\sqrt{\varepsilon_0\mu_0}$ denotes the light velocity in empty space. From (15) it is immediate the conservation law of the electromagnetic energy:

$$\frac{\partial}{\partial t}\left(\frac{\varepsilon_0}{2}E^2 + \frac{1}{2\mu_0}B^2\right) + \vec{\nabla}\cdot\vec{P} = 0, \qquad \vec{P} = \frac{1}{\mu_0}\vec{E}\times\vec{B}, \qquad (16)$$

With $\vec{P}$ the Poynting vector [35].

The equations (15) are invariant under the duality rotations [21, 36–42]:

$$c\vec{B}' = c\vec{B}\cos\alpha - \vec{E}\sin\alpha, \quad \vec{E}' = c\vec{B}\sin\alpha + \vec{E}\cos\alpha, \tag{17}$$

Because the fields (17) also satisfy (15). Here we employ the Lanczos approach to the Noether theorem to show that the invariance of (15) under (17) implies (16). In fact, if we use the complex Riemann-Silberstein vector [18, 42–46] $\vec{F} = c\vec{B} + i\vec{E}, i = \sqrt{-1}$, then the relations (15) are equivalent to:

$$\vec{\nabla} \times \vec{F} + \frac{i}{c}\frac{\partial\vec{F}}{\partial t} = \vec{0}, \tag{18}$$

And the duality rotations (17) represent the change of phase:

$$\vec{F}' = e^{i\alpha}\vec{F}, \quad \alpha = constant. \tag{19}$$

On the other hand, the Maxwell equations (18) can be deduced from the principle [18]:

$$\delta\int_{V_4} L\, d^4x = 0, \quad L = \vec{F}^* \cdot \left(\vec{\nabla} \times \vec{F} + \frac{i}{c}\frac{\partial\vec{F}}{\partial t}\right), \quad \vec{F}^* = c\vec{B} - i\vec{E}, \tag{20}$$

Where $\vec{F}$ and $\vec{F}^*$ are the variational variables. It is clear that L is invariant under the global symmetry (19) with α = constant, then the Lanczos approach to Noether's theorem indicates to calculate L' but now with (19) as an infinitesimal local symmetry because $\alpha(\vec{r},t) \ll 1$ is a new degree of freedom, with its corresponding Euler-Lagrange equation:

$$\frac{\partial}{\partial x}\left(\frac{\partial L'}{\partial\alpha_{,x}}\right) + \frac{\partial}{\partial y}\left(\frac{\partial L'}{\partial\alpha_{,y}}\right) + \frac{\partial}{\partial z}\left(\frac{\partial L'}{\partial\alpha_{,z}}\right) + \frac{\partial}{\partial t}\left(\frac{\partial L'}{\partial\alpha_{,t}}\right) = 0. \tag{21}$$

Therefore:

$$L' \equiv \vec{F}'^* \cdot \left(\vec{\nabla} \times \vec{F}' + \frac{i}{c}\frac{\partial\vec{F}'}{\partial t}\right), \vec{F}' = e^{i\alpha(\vec{r},t)}\vec{F} = (1 + i\alpha)\vec{F}, \quad \alpha \ll 1,$$

$$= L - i\left(\vec{F}^* \times \vec{F}\right)\cdot\vec{\nabla}\alpha - \frac{1}{c}\vec{F}^*\cdot\vec{F}\alpha_{,t}\,, \text{ to first order in } \alpha$$

Thus (21) implies the continuity equation:

$$\frac{\partial}{\partial t}\left(\vec{F}^* \cdot \vec{F}\right) + i\vec{\nabla}\cdot\left(c\vec{F}^* \times \vec{F}\right) = 0, \tag{22}$$

which is equivalent to (16) because $\vec{F}^* \cdot \vec{F} = c^2 B^2 + E^2$ and $c\vec{F}^* \times \vec{F} = -\frac{2i}{\varepsilon_0} \vec{P}$. The process here realized manifests that, in the source-free case, the invariance of Maxwell equations under duality rotations is the underlying symmetry into the conservation law of the electromagnetic energy [21].

8.5 HOMOGENEOUS LINEAR DIFFERENTIAL EQUATION OF SECOND ORDER

Now we consider the differential equation [26]:

$$p(x)y'' + q(x)y' + r(x)y = 0, \tag{23}$$

And we accept to have the solution $y_1(x)$:

$$p\,y_1'' + q\,y_1' + r\,y_1 = 0. \tag{24}$$

It is very known [48] that the Abel-Liouville-Ostrogradski expression [49] for the Wronskian:

$$W \equiv y_1 y_2' - y_2 y_1' = \exp\left(-\int^x \frac{q(\xi)}{p(\xi)} d\xi\right), \tag{25}$$

Allows to obtain other solution for (23):

$$y_2(x) = y_1(x) \int^x \frac{W(\xi)}{\left[y_1(\xi)\right]^2} d\xi. \tag{26}$$

Here we employ a variational approach to construct (26); in fact, it is easy to see that the action [50]:

$$S = \int_{x_1}^{x_2} L(y, y', x) dx = \int_{x_1}^{x_2} \left(y'^2 - \frac{r}{p} y^2\right) \exp\left(\int^x \frac{q}{p} d\xi\right) dx, \tag{27}$$

Implies (23) under the condition $\delta S = 0$, that is, the Euler-Lagrange expression [18] $\frac{d}{dx}\left(\frac{\partial L}{\partial y'}\right) - \frac{\partial L}{\partial y} = 0$ leads to our homogeneous differential equation. Now we show that (27) admits a variational symmetry [32], hence we apply the Noether's theorem in the Lanczos approach to prove that the corresponding conserved quantity gives the relation (26) for the solution $y_2(x)$. In fact, we use y_1 to introduce the infinitesimal transformation:

$$\tilde{y}(x) = y(x) + \varepsilon\, y_1(x), \qquad \varepsilon \ll 1, \tag{28}$$

Where ε is a constant parameter, then the new Lagrangian into (27) is given by:

$$
\begin{aligned}
\tilde{L} &= \left(\tilde{y}'^2 - \frac{r}{p}\tilde{y}^2 \right) \\
&= L + \frac{d}{dx}\left[2\varepsilon\, y\, y_1' \exp\left(\int^x \frac{q}{p} d\xi \right) \right] - \frac{2\varepsilon}{p} y\left(py_1' + qy_1' + ry_1 \right)\exp\left(\int^x \frac{q}{p} d\xi \right), \\
&= L + \frac{d}{dx}\left[2\varepsilon\, y\, y_1' \exp\left(\int^x \frac{q}{p} d\xi \right) \right],
\end{aligned}
\tag{29}
$$

Therefore the Lagrangian is invariant up to divergence, hence (28) is a variational symmetry of the action (27).

To study the constant of motion associated to this symmetry (28) we apply the Noether theorem via the Lanczos technique, that is, into L we employ the transformation (28) but now $\varepsilon(x)$ is a new degree of freedom:

$$\tilde{L} = L + 2\left[\left(y'y_1' - \frac{r}{p} y\, y_1 \right)\varepsilon + y'y_1\, \varepsilon' \right]\exp\left(\int^x \frac{q}{p} d\xi \right), \tag{30}$$

Thus the Euler-Lagrange equation $\dfrac{d}{dx}\left(\dfrac{\partial \tilde{L}}{\partial \varepsilon'} \right) - \dfrac{\partial \tilde{L}}{\partial \varepsilon} = 0$ and (24) imply the relation:

$$\frac{d}{dx}\left[\left(y'y_1 - y\, y_1' \right)\exp\left(\int^x \frac{q}{p} d\xi \right) \right] = 0, \tag{31}$$

Whose integration gives the solution (26). Thus, the Lanczos version of Noether theorem shows that the structure of equation (26) is consequence from a variational symmetry of the action associated to the homogeneous differential equation (23).

We can study the general solution of the second-order linear differential equation:

$$p(x)y'' + q(x)y' + r(x)y = \phi(x), \tag{32}$$

Via an alternative (but equivalent) method to the variation of parameters technique of Newton (Principia)-Bernoulli-Euler-Lagrange [48, 51–53]. We have the Abel-Liouville-Ostrogradski expression (25) for the

Wronskian where y_1 and y_2 are solutions of the corresponding homogeneous equation (23):

$$p(x)y''_k + q(x)y'_k + r(x)y_k = 0, \qquad k = 1, 2, \tag{33}$$

Hence if we know y_1 then from (25) we can to construct the solution (26).

Now we show a procedure to obtain the particular solution y_p verifying:

$$p(x)y''_p + q(x)y'_p + r(x)y_p = \phi(x), \tag{34}$$

And it is not necessary the Lagrange's ansatz; thus the general solution of (32) is given by:

$$y(x) = c_1 y_1(x) + c_2 y_2(x) + y_p(x). \tag{35}$$

If we multiply (33), for $k=1$, by $\frac{1}{Wp}$ and we use (25), it is easy to deduce the relation:

$$r\frac{y_1}{Wp} = -\frac{d}{dx}\left(\frac{y_1'}{W}\right). \tag{36}$$

Similarly, if we multiply (32) by $\frac{\lambda}{p}$, where $\lambda(x)$ is a function to determine, we find that:

$$\left[\frac{r\lambda}{p} - \frac{d}{dx}\left(\frac{\lambda q}{p} - \lambda'\right)\right]y + \frac{d}{dx}\left[\left(\frac{\lambda q}{p} - \lambda'\right)y + \lambda y'\right] = \frac{\lambda}{p}\phi, \tag{37}$$

Whose left side is an exact derivative if we ask:

$$r\frac{\lambda}{p} = \frac{d}{dx}\left(\frac{\lambda q}{p} - \lambda'\right) \equiv -\frac{d}{dx}\left[\frac{1}{W}(W\lambda)'\right], \tag{38}$$

Then its comparison with (36) leads to $\lambda = \frac{y_1}{W}$, and (37) acquires the form:

$$\frac{d}{dx}\left[\frac{y_1^2}{W}\frac{d}{dx}\left(\frac{y}{y_1}\right)\right] = \frac{y_1\phi}{pW}, \tag{39}$$

Therefore, two successive integrations of (39) imply the general solution (35) with y_2 given by (26), and the particular solution:

$$y_p(x) = y_2(x) \int^x \frac{y_1(\xi)\phi(\xi)}{p(\xi)W(\xi)} d\xi - y_1(x) \int^x \frac{y_2(\xi)\phi(\xi)}{p(\xi)W(\xi)} d\xi , \tag{40}$$

In harmony with the variation of parameters method [26, 48, 54]; we consider that our approach justifies the traditional Lagrange's ansatz employed in that method.

We note that the action (27) can be extended to [55]:

$$\overline{S} = \int_{x_1}^{x_2} \left(y'^2 - \frac{r}{p} y^2 + \frac{2\phi}{p} y \right) \exp\left(\int^x \frac{q(\xi)}{p(\xi)} d\xi \right) dx, \tag{41}$$

Such that $\delta\overline{S} = 0$ implies (32).

8.6   CONCLUSIONS

The energy-momentum tensor of the electromagnetic field $T^{\mu\nu} = T^{\nu\mu}$, whose symmetry was required by Planck [56] to secure the relativistic equivalence between mass and energy, has null trace ($T^{\mu}{}_{\mu} = 0$) because the photon being mass less; the photon has spin 1 because the Maxwell spin or possesses two indices [57]. Thus, we have the following quadratic expression in the Faraday tensor [58]:

$$T^{\mu\nu} = -F^{\mu\alpha} F^{\nu}{}_{\alpha} + \frac{1}{4} I_1 g^{\mu\nu} , \qquad I_1 = F^{\alpha\beta} F_{\alpha\beta} , \tag{42}$$

Which is invariant under duality rotations as expected [that is, (42) is not altered if we use the fields (17)], due to the fact that this symmetry is associated with the conservation of the electromagnetic energy. Warwick [59] exposes an interesting story about how Poynting discovered the continuity equation (16). In Section 8.5, we considered the linear differential equation of second order and we showed that our process (without the ansatz of Lagrange) motivates the method of variation of parameters; it must be valuable apply our approach, based in the Lanczos version of Noether's theorem, to differential equations of third and fourth order [26].

KEYWORDS

- complex Riemann-Silberstein vector
- duality rotations
- ignorable variable
- invariance of the action
- Lanczos variational method
- linear differential equation
- Maxwell equations
- Noether's theorem
- variational symmetry

REFERENCES

1. Leibnitz, G., (1669). Dynamic Power and the Laws of Bodily Nature, (published in 1890).
2. Noether, E., (1918). Invariant variation problems, *Nachr. Ges. Wiss. Göttingen, 2,* 235–257.
3. Kosmann-Schwarzbach, Y., (2011). *The Noether Theorems.* Springer, New York.
4. Havas, P., (1973). The connection between conservation laws and invariance groups: Folklore, fiction, and fact. *Acta Physica Austriaca, 38,* 145–167.
5. Bessel-Hagen, E., (1921). Über die erhaltungssätze der elektrodynamik. *Mathematische Annalen, 84,* 258–276.
6. Weitzenböck, R., (1923). *Invariantentheorie.* Groningen: Noordhoff.
7. Neuenschwander, D. E., (1998). *Symmetries, Conservation Laws, and Noether's Theorem, Radiations, Fall,* 12–15.
8. Neuenschwander, D. E., (2011). *Emmy Noether's Wonderful Theorem.* The Johns Hopkins University Press, Baltimore.
9. Havelková, M., (2012). Symmetries of a dynamical system represented by singular Lagrangians. *Comm. in Maths., 20*(1), 23–32.
10. Trautman, A., (1967). Noether's equations and conservation laws. *Commun. Math. Phys., 6*(4), 248–261.
11. Rund, H., (1972). A direct approach to Noether's theorem in the calculus of variations. *Utilitas Math., 2,* 205–214.
12. Sinaceur, M. A., (1990). Dedekind et le programme de Riemann. *Rev. Hist. Sci., 43,* 221–294.
13. Laugwitz, D., (2008). Bernhard Riemann 1826–1866. *Turning Points in the Conception of Mathematics.* Birkhäuser, Boston MA.
14. Byers, N., (1996). *Emmy Noether's Discovery of the Deep Connection Between Symmetries and Conservation Laws.* Proc. Symp. Heritage E. Noether, Bar Ilan University, Tel Aviv, Israel.

15. Brading, K. A., & Brown, H. R., (2003). Symmetries and Noether's theorems. In: Katherine, A. B., & Elena, C., (eds.), *Symmetries in Physics Philosophical Reflections* (pp. 89–109). Cambridge University Press.

16. Lederman, L. M., & Ch, T. H., (2004). *Symmetry and the Beautiful Universe.* Prometheus Books, Amherst, New York, Chaps 3 and 5.

17. Hargittai, M., & Hargittai, I., (2004). *Candid Science IV: Conversations with Famous Physicists.* Imperial College Press, London.

18. Lanczos, C., (1970). *The Variational Principles of Mechanics.* University of Toronto Press, Chapter 11.

19. Lanczos, C., (1973). Emmy Noether and the calculus of variations. *Bull. Inst. Math. and Appl., 9*(8), 253–258.

20. Lam-Estrada, P., López-Bonilla, J., López-Vázquez, R., & Ovando, G., (2014). Lagrangians: Symmetries, gauge identities, and first integrals. *The Sci. Tech. J. of Sci. and Tech., 3*(1), 54–66.

21. López-Bonilla, J., López-Vázquez, R., & Man, T. B., (2014). Maxwell equations and duality rotations. *The SciTech, J. of Sci. and Tech., 3*(2), 20–22.

22. Lam-Estrada, P., López-Bonilla, J., López-Vázquez, R., & Ovando, G., (2015). On the gauge identities and genuine constraints of certain lagrangians. *Prespacetime Journal, 6*(3), 238–246.

23. Lam-Estrada, P., López-Bonilla, J., & López-Vázquez, R., (2014). Lanczos approach to Noether's theorem. *Bull. Soc. for Mathematical Services and Standards (India), 3*(3), 1–4.

24. Lam-Estrada, P., López-Bonilla, J., López-Vázquez, R., Ovando, G., & Vidal-Beltrán, S., (2018). *Noether and Matrix Methods to Construct Local Symmetries of Lagrangians* (Vol. 97, pp. 51–68). World Scientific News.

25. Torres, D. C. G. F., & Rubalcava-García, I., (2017). Variational symmetries as the existence of ignorable coordinates. *European J. Phys., 38*, 025002.

26. Hernández-Galeana, A., López-Bonilla, J., López-Vázquez, R., & Vidal-Beltrán, S., (2018). *Linear Differential Equations of Second, Third, and Fourth Order* (Vol. 105, pp. 225–232). World Scientific News.

27. Neuenschwander, D. E., (1996). *Elegant Connections in Physics: Symmetries, Conservation Laws, and Noether's Theorem.* Soc. of Physics Students Newsletter, Arizona State Univ., Tempe, AZ, USA.

28. Henneaux, M., Teitelboim, C., & Zanelli, J., (1990). Gauge invariance and degree of freedom count. *Nucl. Phys. B, 332*(1), 169–188.

29. Rothe, H. J., & Rothe, K. D., (2010). Classical and quantum dynamics of constrained Hamiltonian systems. *World Scientific, Lecture Notes in Physics, 81*, Singapore.

30. Torres, D. C. G. F., (2014). Point symmetries of the Euler-Lagrange equations. *Rev. Mex. Fís., 60*, 129–135.

31. Bahadur, T. G., Hernández-Galeana, A., & López-Bonilla, J., (2018). *Continuity Equations in Curved Spaces* (Vol. 105, pp. 197–203). World Scientific News.

32. Torres, D. C. G. F., Andrade-Mirón, C., & Bravo-Rojas, R., (2013). Variational symmetries of lagrangians. *Rev. Mex. Fís, E, 59*, 140–147.

33. Routh, E. J., (1877). *Dynamics of Rigid Bodies.* Macmillan.

34. Helmholtz, H. V., (1884). *Journal of Math., 97*, 111.

35. Poynting, J. H., (1884). On the transfer of energy in the electromagnetic field. *Phil. Trans. Roy. Soc. London, 175*, 343–361.

36. Rainich, G. Y., (1925). Electrodynamics in the general relativity theory. *Trans. Amer. Math. Soc., 27*, 106–136.

37. Misner, C. W., & Wheeler, J. A., (1957). Classical physics as geometry. *Ann. of Phys., 2*(6), 525–603.

38. Wheeler, J. A., (1962). *Geometrodynamics*. Academic Press, New York.

39. Witten, L., (1962). A geometric theory of the electromagnetic and gravitational fields. In: *Gravitation: An Introduction to Current Research*. Wiley, New York, Chapter 9.

40. Penney, R., (1964). Duality invariance and Riemannian geometry. *J. Math. Phys., 5*(10), 1431–1437.

41. Torres, D. C. G. F., (1997). Duality rotations in the linearized Einstein theory. *Rev. Mex. Fís., 43*(1), 25–32.

42. Acevedo, M., López-Bonilla, J., & Sánchez-Meraz, M., (2005). *Quaternions, Maxwell Equations and Lorentz Transformations, Apeiron, 12*(4), 371–384.

43. Riemann, B., (1901). In: Weber, H., (ed.), *Die Partiellen Differential-Gleichungen Der Mathematische Physic* (Vol. 2). Vieweg, Braunschweig.

44. Silberstein, L., (1907). *Elektromagnetische Grundgleichungen in bivektorieller behandlung, Ann. Der Physik., 22*, 579–586.

45. Silberstein, L., (1907). Nachtrang zur abhandlung über electromagnetische grundgleichungen in bivektorieller behandlung. *Ann. Der Physik., 24*, 783–784.

46. Bialynicki-Birula, I., (1996). Photon wave function. *Progress in Optics, 36*. E. Wolf, Elsevier, Amsterdam.

47. Hamdan, N., Guerrero-Moreno, I., López-Bonilla, J., & Rosales, L., (2008). On the complex Faraday vector. *The Icfai Univ. J. Phys., 1*(3), 52–56.

48. Ahsan, Z., (2004). *Differential Equations and their Applications*. Prentice-Hall, New Delhi.

49. Elsgotz, L., (1983). *Differential Equations and Variational Calculus*. MIR, Moscow.

50. Lanczos, C., (1934). A new transformation theory of linear canonical equations. *Ann. Der Physik., 20*(5), 653–688.

51. Srinivasan, G., (2007). A note on Lagrange's method of variation of parameters. *Missouri J. Math. Sci., 19*(1), 11–14.

52. Quinn, T., & Rai, S., (2012). Variation of parameters in differential equations. *Primus, 23*(1), 25–44.

53. López-Bonilla, J., Romero-Jiménez, D., & Zaldívar-Sandoval, A., (2016). Variation of parameters method via the Riccati equation. *Prespacetime Journal, 7*(8), 1217–1219.

54. Bahadur, T. G., Domínguez-Pacheco, A., & López-Bonilla, J., (2015). On the linear differential equation of second order. *Prespacetime Journal, 6*(10), 999–1001.

55. López-Bonilla, J., Posadas-Durán, G., & Salas-Torres, O., (2017). Variational principle for py" + qy' + ry = ϕ. *Prespacetime Journal, 8*(2), 226–228.

56. Planck, M., (1908). Bemerkungen zum prinzip der action und reaction in der allgemeinen dynamic. *Phys. Z., 9*, 828–830.

57. Penrose, R., (1999). Spinors in general relativity. *Acta Physica Polonica B, 30*(10), 2979–2987.

58. Synge, J. L., (1965). *Relativity: The Special Theory*. North-Holland, Amsterdam.

59. Warwick, A., (2003). *Masters of Theory: Cambridge and the Raise of Mathematical Physics*. The University of Chicago Press, Chapter 6.

CHAPTER 9

Mathematical Modeling of Elastic-Plastic Transitional Stresses in Human Femur and Tibia Bones Exhibiting Orthotropic Macro Structural Symmetry

SHIVDEV SHAHI and S. B. SINGH

Department of Mathematics, Punjabi University Patiala, Punjab – 147002, India, E-mails: shivdevshahi93@gmail.com (S. Shahi), sbsingh69@yahoo.com (S. B. Singh)

ABSTRACT

In this chapter, elastic-plastic stress distributions in the human femur and tibia bone are calculated analytically. The bone is modeled in the form of a cylinder which exhibits orthotropic macroscopic symmetry. Seth's transition theory has been used to model the elastic-plastic state of stresses. The cylinder so modeled is subjected to external pressure. The results obtained illustrate the stress build up in the bones when there is an external force applied on them, thereby providing insights in the elastic-plastic extensions and their tendency to fracture.

9.1 INTRODUCTION

Humans have evolved in a manner to bear an erect posture. The posture is supported by the bones and muscles of hind limbs which carry the maximum weight of the body. The limbs are primarily comprised of the femur and the tibia bone along with smaller bones of the knee, ankle, and foot. Bones are both anisotropic and heterogeneous in their mechanical properties. Due to various forces to which the bone is subjected, there is a continuous state of stress which not only depends on the intensity and the manner of application of force but also on the mechanical properties of the bone structure. Bundy [1] experimentally demonstrated for the human femoral bone that the mechanical properties vary along the length of the bone which supports the orthotropic elastic behavior. He also showed that the bone is anisotropic but did not measure all of the elastic constants needed to completely characterize the anisotropy. Most investigators examining the anisotropy of bone have assumed it to be transversely isotropic and have measured the five elastic constants needed to characterize such a material [8, 21]. None of these investigators appears to have systematically examined the heterogeneous nature of these properties. The investigation of stress distribution is of prime significance because these results are of much help to orthopedic surgeons and the agencies which work on the design of prosthetic limbs. The stress distribution in the bone implants if available at hand, may help to improve the success rates of surgeries. Pitkin et al. [7] proposed to use titanium as a base material to design the pylon of prosthetic legs. Medical grade Titanium is also used inserted in femur and tibia bones in cases of multiple fractures and in cases when the bones are damaged due to osteoporosis [3, 9]. Bhatnagar et al. [2] studied creep behavior of orthotropic rotating discs having variable thickness (constant, linear, and hyperbolic) using Norton's power law. Gupta and Singh [18] further studied the deformation behavior of a functionally graded disc using similar yielding criteria. These works however did not consider the non-linear character of the transition phase which was defined by B. R. Seth [10–12]. Seth's transition theory is used in this paper to obtain elastic-plastic stresses in human femur bone considering the orthotropic elastic constants obtained using ultrasonic measurement by Buskirk et al. [20]. The concept of generalized strain measures and Seth's transition theory [10, 11] has been applied to find elastic-plastic stresses in problems associated to various structural components [13, 15] by solving the non-linear differential equations at the transition points. All these problems based on the recognition of the transition state as separate state (Figures 9.1 and 9.2).

9.2   GOVERNING EQUATIONS

Consider a hollow circular cylinder of internal and external radii a and b respectively subject to pressure p_o as represented in Figure 9.3.

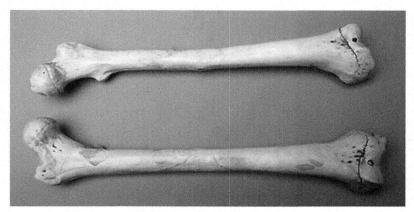

FIGURE 9.1 Human femur bones.

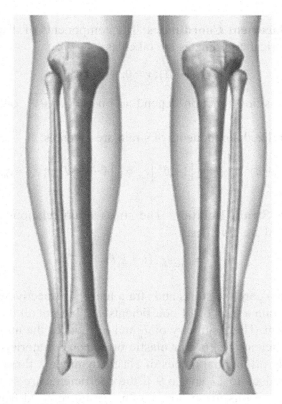

FIGURE 9.2 Human tibia bones.

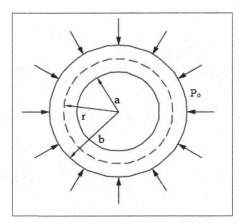

FIGURE 9.3   Geometrical cross-section of a cylinder under external pressure p_o.

1. **Displacement Coordinates**: The components of displacement in cylindrical coordinates are taken as:

$$u = r(1-\beta); v = 0; w = dz \tag{1}$$

where, β is position function depending on $r = \sqrt{x^2 + y^2}$ only and d is a constant.

The generalized components of strain are given as:

$$e_{rr} = \frac{1}{2}\left[1-(r\beta'+\beta)^2\right], e_{\theta\theta} = \frac{1}{2}\left[1-\beta^2\right] e_{zz} = \frac{1}{2}\left[1-(1-d)^2\right], e_{r\theta} = e_{\theta z} = e_{zr} = 0. \tag{2}$$

and $\beta' = d\beta/dr$

2. **Stress-Strain Relation**: The stress-strain relations for isotropic material are given as:

$$T_{ij} = c_{ijkl}e_{kl}, (i,j,k,l = 1,2,3)$$

where, T_{ij} and e_{kl} are the stress and strain tensors respectively. These nine equations contain a total of 81 coefficients e_{ijkl}, but not all the coefficients are independent. The symmetry of T_{ij} and e_{kl} reduces the number of independent coefficients to 36. For elastic orthotropic materials which have three mutually orthogonal planes of elastic symmetry, these independent coefficients reduce to 12 and to 9 if the coefficients are symmetric. The constitutive equations for orthotropic media are referred from Altenbach et al. [1]:

$$\begin{bmatrix} T_{11} \\ T_{22} \\ T_{33} \\ T_{23} \\ T_{31} \\ T_{12} \end{bmatrix} = \begin{bmatrix} c_{11} & c_{12} & c_{13} & 0 & 0 & 0 \\ c_{21} & c_{22} & c_{23} & 0 & 0 & 0 \\ c_{31} & c_{32} & c_{33} & 0 & 0 & 0 \\ 0 & 0 & 0 & c_{44} & 0 & 0 \\ 0 & 0 & 0 & 0 & c_{55} & 0 \\ 0 & 0 & 0 & 0 & 0 & c_{66} \end{bmatrix} \begin{bmatrix} e_{11} \\ e_{22} \\ e_{33} \\ e_{23} \\ e_{31} \\ e_{12} \end{bmatrix}, \tag{3}$$

Substituting Eqn. (2) in Eqn. (3), we get:

$$T_{rr} = \frac{c_{11}}{2}\left[1-\left(r\beta'+\beta\right)^2\right]+\frac{c_{12}}{2}\left(1-\beta^2\right)+\frac{c_{13}}{2}\left(1-(1-d)^2\right);$$

$$T_{\theta\theta} = \frac{c_{21}}{2}\left[1-\left(r\beta'+\beta\right)^2\right]+\frac{c_{22}}{2}\left(1-\beta^2\right)+\frac{c_{23}}{2}\left(1-(1-d)^2\right);$$

$$T_{zz} = \frac{c_{31}}{2}\left[1-\left(r\beta'+\beta\right)^2\right]+\frac{c_{32}}{2}\left(1-\beta^2\right)+\frac{c_{33}}{2}\left(1-(1-d)^2\right);$$

$$T_{r\theta} = T_{\theta z} = T_{zr} = 0 \tag{4}$$

3. **Equation of Equilibrium**: The equations of equilibrium are all satisfied except:

$$\frac{d}{dr}T_{rr} + \frac{T_{rr}-T_{\theta\theta}}{r} = 0 \tag{5}$$

4. **Critical Points or Turning Points**: By substituting Eqn. (4) into Eqn. (5), we gets a non-linear differential equation with respect to b β:

$$\beta\frac{dP}{d\beta} = \frac{[(c_{11}-c_{21})\{1-\beta^n(P+1)^n\}+(c_{12}+c_{22})(1-\beta^n)+(c_{13}-c_{23})\{1-(1-d)^n\}]}{nc_{11}\beta^nP(P+1)^{n-1}} - \frac{c_{12}(1-\beta^n)}{c_{11}(P+1)^{n-1}} - (P+1) \tag{6}$$

where, P is function of β and β is function of r only.

    5.   Transition Points: The transition points of b in Eqn. (6) are $P \to 0$, $P \to -1$ and $P \to \pm\infty$.

The only critical point of interest is $P \to \pm\infty$ which is sufficient for the elastic-plastic state given by Seth [10, 11].

    6.   **Boundary condition:** The boundary conditions of the problem are given by:

$$r = a, \tau_{rr} = 0$$
$$r = b, \tau_{rr} = -p_0 \qquad (7)$$

9.3   PROBLEM SOLUTION

For finding the elastic-plastic stress, the transition function is taken through the principal stress [13, 15] at the transition point $P \to \pm\infty$, we define the transition function ζ as:

$$\zeta = 1 - \frac{nT_{rr}}{(c_{11} + c_{12} + c_{13})} \cong \frac{1}{(c_{11} + c_{12} + c_{13})} \left[c_{11}\beta^n (P+1)^n + c_{12}\beta^n + c_{13}(1-d)^n \right] \qquad (8)$$

where, ζ be the transition function of r only.

Substituting the value of T_{rr} from equation (4) in equation (8)

$$\zeta = \frac{c_{11}}{(c_{11} + c_{12} + c_{13})} \left[\beta^n (P+1)^n + \frac{c_{12}}{c_{11}}\beta^n + \frac{c_{13}}{c_{11}}(1-d)^n \right] \qquad (9)$$

Taking the logarithmic differentiation of Eqn. (9), with respect to r and using Eqn. (6), we get:

$$\frac{d(\log \zeta)}{dr} = \left(\frac{n\beta^n P \left[(P+1)^n + \beta \dfrac{dP}{d\beta}(P+1)^{n-1} + \dfrac{c_{12}}{c_{11}} \right]}{r \left[\beta^n (P+1)^n + \beta^n \dfrac{c_{12}}{c_{11}} + \dfrac{c_{13}}{c_{11}}(1-d)^n \right]} \right) \qquad (10)$$

Taking the asymptotic value of Eqn. (10) as $P \to \pm\infty$ and integrating, we get:

$$\zeta = A_0 r^{-K} \qquad (11)$$

where, A_0 is a constant of integration and $K = c_{11} - c_{21}/c_{11}$. From Eqn. (8) and (11), we have:

$$T_{rr} = \frac{(c_{11} + c_{12} + c_{13})}{n} \left[1 - A_0 r^{-K} \right] \qquad (12)$$

The value of material constant in the transition range is:

$$Y = \frac{1}{n} \frac{\left[c_{11}c_{22}c_{33} - c_{11}c_{23}^2 - c_{22}c_{12}^2 - c_{33}c_{12}^2 + 2c_{23}c_{12}^2 \right]}{\left[c_{22}c_{33} - c_{23}^2 \right]} \qquad (13)$$

where, Y is yield stress in tension.

Using the yield stress in eqn. (12), we get:

$$T_{rr} = \frac{\left(c_{11} + c_{12} + c_{13} \right) Y (c_{22}c_{33} - c_{23}^2) \left[1 - A_0 r^{-K} \right]}{\left[c_{11}c_{22}c_{33} - c_{11}c_{23}^2 - c_{22}c_{12}^2 - c_{33}c_{12}^2 + 2c_{23}c_{12}^2 \right]} \qquad (14)$$

Substituting eqn. (14) in eqn. of equilibrium:

$$T_{\theta\theta} - T_{rr} = \frac{\left(c_{11} + c_{12} + c_{13} \right) Y (c_{22}c_{33} - c_{23}^2) A_0 (c_{11} - c_{21}) r^{-K}}{c_{11} \left[c_{11}c_{22}c_{33} - c_{11}c_{23}^2 - c_{22}c_{12}^2 - c_{33}c_{12}^2 + 2c_{23}c_{12}^2 \right]} \qquad (15)$$

$$T_{\theta\theta} = \frac{\left(c_{11} + c_{12} + c_{13} \right) Y (c_{22}c_{33} - c_{23}^2) \left[1 - (1-K)A_0 r^{-K} \right]}{\left[c_{11}c_{22}c_{33} - c_{11}c_{23}^2 - c_{22}c_{12}^2 - c_{33}c_{12}^2 + 2c_{23}c_{12}^2 \right]} \qquad (16)$$

Using boundary conditions of eqn. (7) in eqn. (14) we get:

$$A_0 = a^K \qquad (17)$$

and

$$P_o = \frac{\left(c_{11} + c_{12} + c_{13} \right) . Y . (c_{22}c_{33} - c_{23}^2) \left[\left(\dfrac{a}{b} \right)^K - 1 \right]}{\left[c_{11}c_{22}c_{33} - c_{11}c_{23}^2 - c_{22}c_{12}^2 - c_{33}c_{12}^2 + 2c_{23}c_{12}^2 \right]} \qquad (18)$$

Substituting eqn. (17) in the principal stresses and principal stress difference, we get:

$$T_{rr} = \frac{\left(c_{11} + c_{12} + c_{13} \right) Y (c_{22}c_{33} - c_{23}^2) \left[1 - \left(\dfrac{a}{r} \right)^K \right]}{\left[c_{11}c_{22}c_{33} - c_{11}c_{23}^2 - c_{22}c_{12}^2 - c_{33}c_{12}^2 + 2c_{23}c_{12}^2 \right]} \qquad (19)$$

$$T_{\theta\theta} - T_{rr} = \frac{(c_{11} + c_{12} + c_{13})Y(c_{22}c_{33} - c_{23}^2)(c_{11} - c_{21})\left(\dfrac{a}{r}\right)^K}{c_{11}\left[c_{11}c_{22}c_{33} - c_{11}c_{23}^2 - c_{22}c_{12}^2 - c_{33}c_{12}^2 + 2c_{23}c_{12}^2\right]} \tag{20}$$

$$T_{\theta\theta} = \frac{(c_{11} + c_{12} + c_{13})Y(c_{22}c_{33} - c_{23}^2)\left[1 - (1-K)\left(\dfrac{a}{r}\right)^K\right]}{\left[c_{11}c_{22}c_{33} - c_{11}c_{23}^2 - c_{22}c_{12}^2 - c_{33}c_{12}^2 + 2c_{23}c_{12}^2\right]} \tag{21}$$

$$T_{zz} = \frac{c_{31}(T_{\theta\theta} + T_{rr})}{c_{11} + c_{21}} + \left[c_{32} - \frac{c_{31}(c_{12} + c_{22})}{c_{11} + c_{21}}\right]\frac{1}{n} + \left[c_{33} - \frac{c_{31}(c_{13} + c_{23})}{c_{11} + c_{21}}\right]e_{zz} \tag{22}$$

where, $e_{zz} = \dfrac{1}{n}\left[1 - (1-d)^n\right]$ and e_{zz} is obtained by considering the cylinder as closed ended:

$$2\pi \int_a^b rT_{zz}dr = \pi b^2 p$$

and

$$e_{zz} = \frac{\left[\dfrac{b^2 p}{(b^2 - a^2)}\left(1 - \dfrac{2c_{31}}{c_{11} + c_{21}}\right) - \left[c_{32} - \dfrac{c_{31}(c_{12} + c_{22})}{c_{11} + c_{21}}\right]\dfrac{1}{n}\right]}{\left[c_{33} - \dfrac{c_{31}(c_{13} + c_{23})}{c_{11} + c_{21}}\right]} \tag{23}$$

Substituting the value of e_{zz} and the values of T_{rr} and $T_{\theta\theta}$ from eqn. (22) and (23) we get:

$$T_{zz} = \frac{c_{31}}{c_{11} + c_{21}}Y\frac{(c_{11} + c_{12} + c_{13})(c_{22}c_{33} - c_{23}^2)\left[2 - 2\left(\dfrac{a}{r}\right)^K + K\left(\dfrac{a}{r}\right)^K\right]}{\left[c_{11}c_{22}c_{33} - c_{11}c_{23}^2 - c_{22}c_{12}^2 - c_{33}c_{12}^2 + 2c_{23}c_{12}^2\right]} + \frac{b^2 p}{(b^2 - a^2)}\left(1 - \frac{2c_{31}}{c_{11} + c_{21}}\right) \tag{24}$$

Initial Yielding: $|T_{\theta\theta} - T_{rr}|$ is maximum at r = b which clearly shows that the yielding will take place at the outer surface. Hence we have:

$$|T_{\theta\theta} - T_{rr}| = \frac{(c_{11} + c_{12} + c_{13})Y(c_{22}c_{33} - c_{23}^2)(c_{11} - c_{21})\left(\dfrac{a}{b}\right)^K}{c_{11}\left[c_{11}c_{22}c_{33} - c_{11}c_{23}^2 - c_{22}c_{12}^2 - c_{33}c_{12}^2 + 2c_{23}c_{12}^2\right]} \cong Y* \text{ (Yielding)} \tag{25}$$

Substituting the value of Y in terms of $Y*$ in eqn (19), (21) and (24) the transitional stresses are given by:

$$T_{rr} = \frac{1}{K}\left(\frac{b}{a}\right)^K Y*\left[1-\left(\frac{a}{r}\right)^K\right]$$
(26)

$$T_{\theta\theta} = \frac{1}{K}\left(\frac{b}{a}\right)^K Y*\left[1-(1-K)\left(\frac{a}{r}\right)^K\right]$$
(27)

$$T_{zz} = \frac{c_{31}}{c_{11}+c_{21}}Y*\frac{1}{K}\left(\frac{b}{a}\right)^K\left[2-2\left(\frac{a}{r}\right)^K+K\left(\frac{a}{r}\right)^K\right]+\frac{b^2 p_i}{(b^2-a^2)}\left(1-\frac{2c_{31}}{c_{11}+c_{21}}\right)$$
(28)

The pressure at initial yielding is calculated using eqn. (18) and (25)

$$p_i = \frac{1}{K}\left(\frac{b}{a}\right)^K Y*\left[\left(\frac{a}{b}\right)^K-1\right]$$
(29)

Converting into non dimensional components:

$R=r/a$, $R_0=b/a$, $\sigma_{rr}=T_{rr}/Y*$, $\sigma_{\theta\theta}=T_{\theta\theta}/Y*$, $\sigma_{zz}=T_{zz}/Y*$ and $P_i=p_i/Y*$

$$\sigma_{rr} = \frac{1}{K}R_0^K\left[1-R^{-K}\right]$$
(30)

$$\sigma_{\theta\theta} = \frac{1}{K}R_0^K\left[1-(1-K)R^{-K}\right]$$
(31)

$$\sigma_{zz} = \frac{c_{31}}{c_{11}+c_{21}}\frac{1}{K}R_0^K\left[2-2R^{-K}+KR^{-K}\right]+\frac{p_i}{(1-R_0^{-2})}\left(1-\frac{2c_{31}}{c_{11}+c_{21}}\right)$$
(32)

$$P_i = \frac{1}{K}R_0^K\left[R_0^{-K}-1\right]$$
(33)

For fully-plastic case, $c_{11}=c_{13}=-c_{12}$, $c_{23}=c_{21}=-c_{22}$ stresses and pressure are calculated as follows [12]:

Where $R=r/b$, $R_0=b/a$, $\sigma_{rr}=T_{rr}/Y^*$, $\sigma_{\theta\theta}=T_{\theta\theta}/Y^*$, $K_1=(c_{11}-c_{22})/c_{11}$ and $P_f=p_f/Y^*$

$$\sigma_{rr} = \frac{1}{K_1}R_0^{K_1}\left[1-R^{-K_1}\right]$$
(34)

$$\sigma_{\theta\theta} = \frac{1}{K_1} R_0^{K_1} \left[1 - (1 - K_1) R^{-K_1} \right] \tag{35}$$

$$\sigma_{zz} = \frac{c_{33}}{c_{11} + c_{21}} \frac{1}{K_1} R_0^{K_1} \left[2 - 2R^{-K_1} + K_1 R^{-K_1} \right] + \frac{p_f}{(1 - R_0^{-2})} \left(1 - \frac{2c_{33}}{c_{11} + c_{21}} \right) \tag{36}$$

$$P_f = \frac{1}{K_1} R_0^{K_1} \left[R_0^{-K_1} - 1 \right] \tag{37}$$

9.4   NUMERICAL RESULTS AND DISCUSSIONS

The above investigations elaborate the initial yielding and fully plastic state of a cylindrical structure subjected to external pressure. The cases of two cylinders were considered, first is the Femur bone and second is the tibia bone; both exhibiting orthotropic material behavior.

In Figure 9.4, curves are plotted for external pressure at initial yielding state and radii ratio $R_0 = b/a$ of the cylinders. The graph has been plotted for considerably thick cylinders. The curves show how the walls of the cylinders yield when various ratios of radial distances were considered. It is observed that the femur bone and the tibia bone yield in a very similar manner when the pressure is applied at the external surface, for the model with considerable thickness, however, the pressures required to yield femur bone were higher as compared to tibia bone.

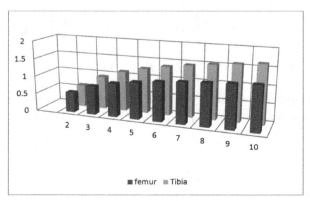

FIGURE 9.4   The pressure at initial yielding in the cylinder.

The radial, circumferential, and axial stresses which were calculated at initial yielding are plotted in Figures 9.5–9.7 and fully plastic state are plotted in Figures 9.8–9.10; along the radii ratio R=r/a.

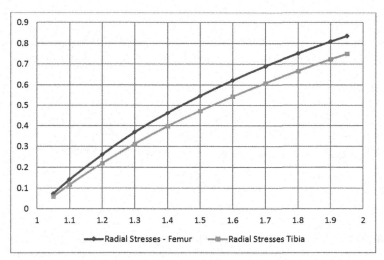

FIGURE 9.5 Radial stresses at initial yielding along the radius ratio r/a.

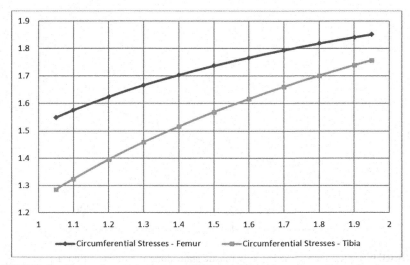

FIGURE 9.6 Circumferential stresses at initial yielding along the radius ratio r/a.

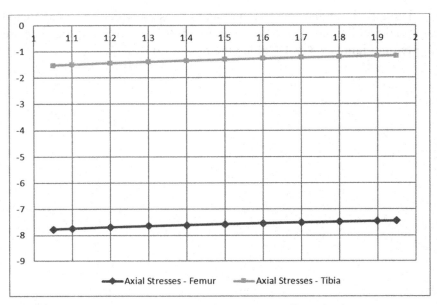

FIGURE 9.7   Axial stresses at initial yielding along the radius ratio r/a.

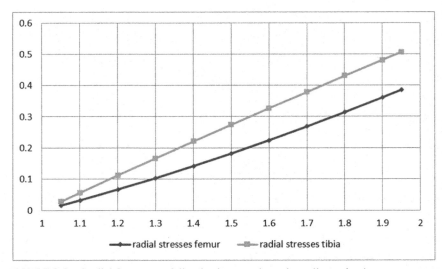

FIGURE 9.8   Radial Stresses at fully plastic state along the radius ratio r/a.

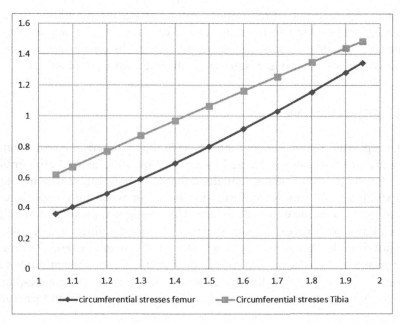

FIGURE 9.9   Circumferential stresses at fully plastic state along the radius ratio r/a.

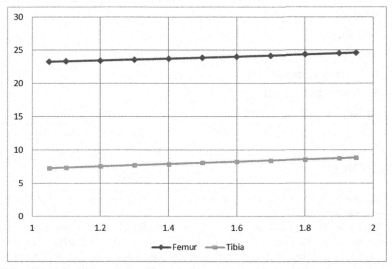

FIGURE 9.10   Axial stresses at fully plastic state along the radius ratio r/a.

It is observed that at the initial yielding stage, the radial and circum-ferential stress concentrations increase with increase in thickness of the bones. However the axial stresses did not very much on application of external pressure. Similar results were observed for fully plastic state.

9.5 CONCLUSIONS

Elastic-plastic stress concentrations have been determined in Human Femur and tibia bones subjected to external pressure using Seth's transi-tion theory. It is sufficient to conclude that both bone structures are likely to exhibit similar kind of deformation behavior when they experience a radial pressure at the external surface. The elastic and plastic limits of the bones have been revealed with their stress withstanding capacity where the initial yielding and fully plastic state has been calculated. The transi-tion phase is also discussed in which the varying stress concentrations are observed. Therefore, it is highly recommended to study the elastic-plastic transition phase to get deeper insights of bone deformations and fractures.

KEYWORDS

- bones
- cylinder
- femur
- orthotropic materials
- Seth's transition theory
- tibia

REFERENCES

1. Altenbach, H., Altenbach, J., & Kissing, W., (2004). *Mechanics of Composite Structural Elements.* Springer-Verlag.
2. Bhatnagar, N. S., Kulkarni, P. S., & Arya, V. K., (1986). Steady state creep of orthotropic rotating discs of variable thickness. *Nucl. Eng. Des., 91*(2), 121–141.
3. Biswas, J. K., Rana, M., Majumder, S., Karmakar, S. K., & Roy, C. A., (2018). Effect of two-level pedicle-screw fixation with different rod materials on lumbar spine: A finite element study. *Journal of Orthopaedic Science, 23*(2), 258–265.

4. Bundy, K. J., (1974). *Experimental Studies of the Non-Uniformity and Anisotropy of Human Compact Bone*. PhD dissertation. Stanford University.

5. Ledbetter, H., Fortunko, C., & Heyliger, P., (1995). Orthotropic elastic constants of a boron-aluminum fiber-reinforced composite: An acoustic-resonance-spectroscopy study. *Journal of Applied Physics, 78*, 1542.

6. Ma, B. M., (1961). Creep analysis of rotating solid discs with variable thickness and temperature. *J. Frankl. Inst., 271*(1), 40–55.

7. Pitkin, M., Raykhtsaum, G., Pilling, J., Galibin, O. V., Protasov, M. V., Chihovskaya, J. V., Belyaeva, I. G., et al., (2007). Porous composite prosthetic pylon for integration with skin and bone. *Journal of Rehabilitation Research and Development, 44*(5), 723–738.

8. Reilly, D. T., & Burstein, A. H., (1975). The elastic and ultimate properties of compact bone tissue. *Journal of Biomechanics, 8*, 393–405.

9. Roy, S., Khutia, N., Das, D., Das, M., Balla, V. K., Bandyopadhyay, A., & Chowdhury, A. R., (2016). Understanding compressive deformation behavior of porous Ti using finite element analysis. *Materials Science and Engineering: C, 64*, 436–443.

10. Seth, B. R., (1962). Transition theory of elastic-plastic deformation, creep, and relaxation. *Nature, 195*, 896, 897.

11. Seth, B. R., (1966). Measure concept in mechanics. *International Journal of Non-Linear Mechanics, 1*(1), 35–40.

12. Seth, B. R., (1972). Yield conditions in plasticity. *Arch. Mech. Strus., 24*(5), 769–776.

13. Shahi, S., Singh, S. B., & Thakur, P., (2019). Modeling creep parameter in rotating discs with rigid shaft exhibiting transversely isotropic and isotropic material behavior. *Journal of Emerging Technologies and Innovative Research, 6*(1), 387–395.

14. Singh, S. B., & Ray, S., (2001). Steady-state creep behavior in an isotropic functionally graded material rotating disc of Al-SiC composite. *Metall. Trans., 32A*(7), 1679–1685.

15. Thakur, P., Shahi, S., Gupta, N., & Singh, S. B., (2017). Effect of mechanical load and thickness profile on creep in a rotating disc by using Seth's transition theory. *AIP Conf. Proc., 1859*(1), 020024.

16. Tresca, H., (1868). *M'emoire Sur I Ecoloment Descrops Solids.' M'eoirepresente's Par Divers Savents, 18*, 733–799.

17. Uyaner, M., Akdemir, A., Erim, S., & Avci, A., (2000). Plastic zones in a transversely isotropic solid cylinder containing a ring-shaped crack. *International Journal of Fracture, 106*, 161–175.

18. Gupta, V., & Singh, S. B., (2016). Mathematical modeling of creep in a functionally graded rotating disc with varying thickness. *Regen. Eng. Transl. Med., 2*, 126–140.

19. Von, M. R., (1913). Mechanics of solids in the plastically deformable state. *NASA Tech Memo, 88488*, 1986.

20. Van, B. W. C., Stephen, C., & Ward, R. N., (1981). Ultrasonic measurement of orthotropic elastic constants of bovine femoral Bone. *Journal of Biomechanical Engineering, 103*(2), 67–72.

21. Yoon, H. S., & Katz, J. L., (1976). Ultrasonic wave propagation in human cortical bone: I. theoretical considerations for hexagonal symmetry. *Journal of Biomechanics, 9*, 407–412.

CHAPTER 10

Lorentz Transformations, Dirac Matrices, and 3-Rotations via Quaternions

J. MORALES,[1] G. OVANDO,[1] J. LÓPEZ-BONILLA,[2] and R. LÓPEZ-VÁZQUEZ[2]

[1]*Area of Physics, Autonomous Metropolitan University, Azcapotzalco, St. Paul Ave. 180, CP 02200, Mexico City, Mexico*

[2]*Superior School of Electrical and Mechanical Engineering, National Polytechnic Institute, Building 4, Lindavista CP 07738, Mexico City, Mexico, E-mail: jlopezb@ipn.mx (J. López-Bonilla)*

ABSTRACT

We study how to generate orthogonal 4×4-matrices using a quaternionic triple product, which leads in a natural manner to Dirac matrices and the analysis of rotations in three and four dimensions.

10.1 INTRODUCTION

The quaternionic units obey the algebra [1–5]:

$$\mathbf{I}^2 = \mathbf{J}^2 = \mathbf{K}^2 = -1, \quad \mathbf{IJK} = -1, \tag{1}$$

Which allows realize the product:

$$\tilde{\mathbf{F}} = \mathbf{p}\,\mathbf{F}, \tag{2a}$$

With:

$$\mathbf{F} = F_1\,\mathbf{I} + F_2\,\mathbf{J} + F_3\,\mathbf{K} + F_4\,, \tag{2b}$$

$$\mathbf{p} = p_1 \mathbf{I} + p_2 \mathbf{J} + p_3 \mathbf{K} + p_4, \tag{2c}$$

Then (2a) acquires the matrix form [6]:

$$\begin{pmatrix} \tilde{F_1} \\ \tilde{F_2} \\ \tilde{F_3} \\ \tilde{F_4} \end{pmatrix} = \begin{pmatrix} p_4 & -p_3 & p_2 & p_1 \\ p_3 & p_4 & -p_1 & p_2 \\ -p_2 & p_1 & p_4 & p_3 \\ -p_1 & -p_2 & -p_3 & p_4 \end{pmatrix} \begin{pmatrix} F_1 \\ F_2 \\ F_3 \\ F_4 \end{pmatrix} \equiv P \begin{pmatrix} F_1 \\ F_2 \\ F_3 \\ F_4 \end{pmatrix} \tag{3}$$

The square of the magnitude of $\mathbf{F}$ is defined by:

$$|\mathbf{F}|^2 = \mathbf{F}\overline{\mathbf{F}} = \overline{\mathbf{F}}\mathbf{F} = F_1^2 + F_2^2 + F_3^2 + F_4^2, \tag{4a}$$

Such that:

$$\overline{\mathbf{F}} = -F_1 \mathbf{I} - F_2 \mathbf{J} - F_3 \mathbf{K} + F_4, \tag{4b}$$

And we note that for any $\mathbf{A}$ and $\mathbf{B}$:

$$\overline{\mathbf{AB}} = \overline{\mathbf{B}}\,\overline{\mathbf{A}}. \tag{4c}$$

Thus from (2.a, 3, 4.a, c):

$$|\tilde{\mathbf{F}}|^2 = \mathbf{p}|\mathbf{F}|^2\,\overline{\mathbf{p}} = \mathbf{p}\overline{\mathbf{p}}|\mathbf{F}|^2, \quad \det P = (\mathbf{p}\overline{\mathbf{p}})^2, \tag{5a}$$

Therefore if $\mathbf{p}$ is unitary:

$$\mathbf{p}\overline{\mathbf{p}} = p_1^2 + p_2^2 + p_3^2 + p_4^2 = 1, \quad \det P = 1, \tag{5b}$$

Then the magnitude of $\mathbf{F}$ is constant:

$$\tilde{F_1}^2 + \tilde{F_2}^2 + \tilde{F_3}^2 + \tilde{F_4}^2 = F_1^2 + F_2^2 + F_3^2 + F_4^2, \tag{6a}$$

Which is equivalent to:

$$\begin{pmatrix} \tilde{F_1} & \tilde{F_2} & \tilde{F_3} & \tilde{F_4} \end{pmatrix} \begin{pmatrix} \tilde{F_1} \\ \tilde{F_2} \\ \tilde{F_3} \\ \tilde{F_4} \end{pmatrix} = \begin{pmatrix} F_1 & F_2 & F_3 & F_4 \end{pmatrix} \begin{pmatrix} F_1 \\ F_2 \\ F_3 \\ F_4 \end{pmatrix}, \tag{6b}$$

Hence (3) and (6b) imply the orthogonal character of the transformation:

$$P^T P = I_{4 \times 4}.$$ (7)

Similarly, the product:

$$\tilde{\mathbf{F}} = \mathbf{F} \mathbf{q},$$ (8a)

Has the matrix representation:

$$\begin{pmatrix} \tilde{F}_1 \\ \tilde{F}_2 \\ \tilde{F}_3 \\ \tilde{F}_4 \end{pmatrix} = \begin{pmatrix} q_4 & q_3 & -q_2 & q_1 \\ -q_3 & q_4 & q_1 & q_2 \\ q_2 & -q_1 & q_4 & q_3 \\ -q_1 & -q_2 & -q_3 & q_4 \end{pmatrix} \begin{pmatrix} F_1 \\ F_2 \\ F_3 \\ F_4 \end{pmatrix} \equiv Q \begin{pmatrix} F_1 \\ F_2 \\ F_3 \\ F_4 \end{pmatrix}$$ (8b)

And (6a) is valid if $\mathbf{q}$ is unitary, then Q is orthogonal with det $Q = 1$
The cases (2a) and (8a) can be unified via a triple quaternionic product:

$$\tilde{\mathbf{F}} = \mathbf{p} \mathbf{F} \mathbf{q},$$ (9a)

That is:

$$\begin{pmatrix} \tilde{F}_1 \\ \tilde{F}_2 \\ \tilde{F}_3 \\ \tilde{F}_4 \end{pmatrix} = D_{4 \times 4} \begin{pmatrix} F_1 \\ F_2 \\ F_3 \\ F_4 \end{pmatrix}, \qquad D = PQ,$$ (9b)

Thus $\left| \tilde{\mathbf{F}} \right|^2 = (\mathbf{p}\bar{\mathbf{p}})(\mathbf{q}\bar{\mathbf{q}}) \left| \mathbf{F} \right|^2$ and (6a) is verified for $\left| \mathbf{p} \right| = \left| \mathbf{q} \right| = 1$ with P and Q orthogonal matrices, therefore:

$$D^T D = I, \qquad \det D = 1.$$ (9c)

Now we shall realize applications of (9.a, b, c): The Section 10.2 shows that D reproduces the sixteen Dirac matrices [7] if $\mathbf{p}$ and $\mathbf{q}$ coincide with the quaternionic units. The Sec 3 considers the case of special relativity because D generates Lorentz transformations [8–10] if $\mathbf{q} = \bar{\mathbf{p}}^*$. The Section 10.4 is dedicated to 3-rotations when $\mathbf{p}$ is real and unitary, that is, $\left| \mathbf{p} \right| = 1$, $\mathbf{p} = \mathbf{p}^*$ and $\mathbf{q} = \bar{\mathbf{p}}$.

10.2   DIRAC MATRICES

In relativistic quantum mechanics are important the following sixteen 4×4-matrices [7]:

$$I, \quad \gamma^0 = \begin{pmatrix} I & 0 \\ 0 & -I \end{pmatrix}, \quad \gamma^5 = \begin{pmatrix} 0 & I \\ I & 0 \end{pmatrix}, \quad \gamma^0\gamma^5 = \begin{pmatrix} 0 & I \\ -I & 0 \end{pmatrix}, \quad \gamma^r = \begin{pmatrix} 0 & \sigma_r \\ -\sigma_r & 0 \end{pmatrix}, \quad (10)$$

$$r = 1, 2, 3$$

$$\sigma^{0r} = -\sigma^{r0} = i\begin{pmatrix} 0 & \sigma_r \\ \sigma_r & 0 \end{pmatrix}, \sigma^{jk} = -\sigma^{kj} = \begin{pmatrix} \sigma_i & 0 \\ 0 & \sigma_i \end{pmatrix} \quad (jkl) \quad \text{is a cyclic}$$

permutation of (123):

With the Cayley-Sylvester-Pauli matrices [1, 11–13]:

$$\sigma_1 = \begin{pmatrix} 0 & 1 \\ 1 & 0 \end{pmatrix}, \qquad \sigma_2 = \begin{pmatrix} 0 & -i \\ i & 0 \end{pmatrix}, \qquad \sigma_3 = \begin{pmatrix} 1 & 0 \\ 0 & -1 \end{pmatrix}, \quad i = \sqrt{-1}. \quad (11)$$

Now in Eq. (9a), we can select to $\mathbf{p}$ and $\mathbf{q}$ as the quaternionic units, for example, if $\mathbf{p} = 1$, $\mathbf{q} = \mathbf{K}$, then from Eq. (9b) we obtain that $D = i\sigma^{31}$; similarly, if $\mathbf{p} = \mathbf{I}$, $\mathbf{q} = \mathbf{J}$, therefore $D = -\gamma^1\gamma^5$, etc. Thus, this process allows construct the table [14]:

$\mathbf{p} \backslash \mathbf{q}$	1	$\mathbf{I}$	$\mathbf{J}$	$\mathbf{K}$	
1	I	γ_1	$-\gamma^3$	$i\sigma^{31}$	
$\mathbf{I}$	σ^{02}	$-\gamma^3\gamma^5$	$-\gamma^1\gamma^5$	$-\gamma^5$	(12)
$\mathbf{J}$	$\gamma^0\gamma^5$	σ^{32}	σ^{12}	$i\gamma^2$	
$\mathbf{K}$	$-i\gamma^2\gamma^5$	$i\sigma^{03}$	$i\sigma^{01}$	γ^0	

If W denotes any matrix into this table, then is simple prove the properties:

$$W^{-1} = W^T, \quad W^* = W, \quad W^T = \pm W, \quad (13)$$

That is, all matrices are real, orthogonal, and symmetric or anti-symmetric.

10.3   LORENTZ TRANSFORMATIONS

In Minkowski space time [8] the linear Lorentz transformations connect the coordinates of two reference frames in uniform relative motion [c is the velocity of light in vacuum]:

$$\begin{pmatrix} i\tilde{x} \\ i\tilde{y} \\ i\tilde{z} \\ c\tilde{t} \end{pmatrix} = D \begin{pmatrix} ix \\ iy \\ iz \\ ct \end{pmatrix}, \tag{14a}$$

Which in geometrical terms corresponds to 4-dimensional rotation:

$$\left(c\tilde{t}\right)^2 - \tilde{x}^2 - \tilde{y}^2 - \tilde{z}^2 = \left(ct\right)^2 - x^2 - y^2 - z^2, \tag{14b}$$

In according with the postulates of special relativity.
If the quaternion (2b) is selected as:

$$\mathbf{F} = ix\,\mathbf{I} + iy\,\mathbf{J} + iz\,\mathbf{K} + ct, \tag{15}$$

Then (6a) and (9b) reproduce (14. a, b), and (9a) implies:

$$i\tilde{x}\,\mathbf{I} + i\tilde{y}\,\mathbf{J} + i\tilde{z}\,\mathbf{K} + c\tilde{t} = \mathbf{p}\left(ix\,\mathbf{I} + iy\,\mathbf{J} + iz\,\mathbf{K} + ct\right)\mathbf{q}, \tag{16a}$$

where we can apply the operations * and (4. b, c) to obtain:

$$i\tilde{x}\,\mathbf{I} + i\tilde{y}\,\mathbf{J} + i\tilde{z}\,\mathbf{K} + c\tilde{t} = \mathbf{p}\left(ix\,\mathbf{I} + iy\,\mathbf{J} + iz\,\mathbf{K} + ct\right)\mathbf{q}, \tag{16b}$$

Whose comparison with (16a) gives $\overline{\mathbf{q}}^* = \mathbf{p}$, that is:

$$\mathbf{q} = \overline{\mathbf{p}}^*, \left|\mathbf{p}\right| = 1 \tag{17}$$

Thus (9a) acquires the structure [8, 15–26]:

$$\tilde{\mathbf{F}} = \mathbf{p}\,\mathbf{F}\,\overline{\mathbf{p}}^*, \mathbf{p}\overline{\mathbf{p}} = 1, \tag{18}$$

Then with (15) we can generate Lorentz transformations verifying (9c). For example, if we select:

$$\mathbf{p} = -i\operatorname{senh}\left(\frac{\tau}{2}\right)\mathbf{K} + \cosh\left(\frac{\tau}{2}\right), \tag{19a}$$

We obtain the known relations:

$$\tilde{x} = x, \quad \tilde{y} = y, \quad \tilde{z} = \gamma(z - vt), \quad \tilde{t} = \gamma\left(t - \frac{v}{c^2}z\right), \gamma = \frac{1}{\sqrt{1 - \frac{v^2}{c^2}}}, \tanh \tau = \frac{v}{c} < 1, \tag{19b}$$

Connecting to two observers with motion in the z-direction and relative velocity v.

10.4   ROTATIONS IN THREE DIMENSIONS

Here we consider Lorentz transformations such that:

$$\tilde{t} = t, \tag{20}$$

Which corresponds to 3-rotations. If we apply (20) into (16a), then it is possible to eliminate t when $\mathbf{pq} = 1$, thus from (5b) and (17):

$$\mathbf{p} = \mathbf{p}^*, \tag{21}$$

That is, the Lorentz matrices generate spatial rotations if in (9a) $\mathbf{p}$ is unitary and real. Then (14a) gives the expressions:

$$\begin{pmatrix} \tilde{x} \\ \tilde{y} \\ \tilde{z} \end{pmatrix} = R \begin{pmatrix} x \\ y \\ z \end{pmatrix}, \quad RR^T = I, \quad \det R = 1, \tilde{x}^2 + \tilde{y}^2 + \tilde{z}^2 = x^2 + y^2 + z^2. \tag{22a}$$

From (9a), (17) and (21) is immediate to deduce the following structure of R reported in the literature [24, 27–34]:

$$R = \begin{pmatrix} 1 - 2\left(p_2^2 + p_3^2\right) & 2(p_1p_2 - p_3p_4) & 2\left(p_1p_3 + p_2p_4\right) \\ 2(p_1p_2 + p_3p_4) & 1 - 2\left(p_1^2 + p_3^2\right) & 2(p_2p_3 - p_1p_4) \\ 2(p_1p_3 - p_2p_4) & 2\left(p_1p_4 + p_2p_3\right) & 1 - 2\left(p_1^2 + p_2^2\right) \end{pmatrix}. \tag{22b}$$

KEYWORDS

- **Dirac matrices**
- **Lorentz transformations**
- **Minkowski space-time**
- **quaternions**
- **relative velocity**
- **spatial rotations**

REFERENCES

1. Kronsbein, J., (1967). Kinematics-quaternions-spinors-and Pauli's spin matrices. *Am. J. Phys., 35*(4), 335–342.
2. Van, D. W. B. L., (1976). Hamilton's discovery of quaternions. *Math. Mag., 49*(5), 227–234.
3. McAllister, L. B., (1989). A quick introduction to quaternions. *Pi Mu Epsilon J., 9,* 23–25.
4. Penrose, R., (2004). *The Road to Reality.* Jonathan Cape, London.
5. Familton, J. C., (2015). *Quaternions: A History of Complex Noncommutative Rotation Groups in Theoretical Physics.* PhD Thesis, Columbia University.
6. Jafari, M., Mortazaasl, H., & Yayli, Y., (2011). De Moivre's formula for matrices of quaternions. *J. of Algebra. Number Theory and Appls., 21*(1), 57–67.
7. Leite-Lopes, J., *Introduction to Quantum Electrodynamics.* Trillas, Mexico.
8. Synge, J. L., (1965). *Relativity: The Special Theory.* North-Holland, Amsterdam.
9. López-Bonilla, J., Morales, J., & Ovando, G., (2002). On the homogeneous Lorentz transformations. *Bull. Allahabad Math. Soc., 17,* 53–58.
10. Carvajal, B., Galaz, M., & López-Bonilla, J., (2007). On the Lorentz matrix in terms of Infeld-van der waerden symbols. *Scientia Magna, 3*(3), 82–84.
11. Cayley, A., (1858). A memoir on the theory of matrices. *London Phil. Trans., 148,* 17–37.
12. Sylvester, J., (1884). *On Quaternions, Nonions, and Sedenions* (Vol. 3, pp. 7–9). Johns Hopkins Circ.
13. Pauli, W., (1927). On the quantum mechanics of magnetic electrons, *Zeits. Für. Physik., 43,* 601–623.
14. López-Bonilla, J., Rosales-Roldán, L., & Zúñiga-Segundo, A., (2009). Dirac matrices via quaternions. *J. Sci. Res., 53,* 253–255.
15. Silberstein, L., (1912). Quaternionic form of relativity. *Phil. Mag., 23*(137), 790–809.
16. Cailler, C., (1917). On some formulas of the theory of relativity, *Archs. Sci. Phys. Nat. Genéve Ser. IV, 44,* 237–255.
17. Dirac, P., (1945). Applications of quaternions to Lorentz transformations. *Proc. Roy. Irish Acad. A, 50*(16), 261–270.

18. Conway, A. W., (1947). Applications of quaternions to rotations in hyperbolic space of four dimensions. *Proc. Roy. Soc. London A, 191*, 137–145.

19. Synge, J. L., (1964). Quaternions, Lorentz transformations and the Conway-Dirac-Eddington matrices. *Comm. Dublin Inst. Adv. Stud. Ser. A*, No. 15.

20. Lanczos, C., (1970). *The Variational Principles of Mechanics*. University of Toronto Press, Canada.

21. Girard, P. R., (1984). The quaternion group and modern physics. *Eur. J. Phys., 5*(1), 25–32.

22. De, L. S., (1996). Quaternions and special relativity. *J. Math. Phys., 37*(6), 2955–2968.

23. Acevedo, M., López-Bonilla, J., & Sánchez, M., (2005). Quaternions, Maxwell equations, and Lorentz transformations. *Apeiron, 12*(4), 371–384.

24. Guerrero-Moreno, I., López-Bonilla, J., & Rosales-Roldán, L., (2008). Rotations in three and four dimensions via 2×2 complex matrices and quaternion's. *The Icfai Univ. J. Phys., 1*(2), 7–13.

25. Carvajal, B., Guerrero-Moreno, I., & López-Bonilla, J., (2009). Matrix and quaternionic approaches to Lorentz transformations. *J. Vect. Rel., 4*(2), 82–85.

26. Guerrero-Moreno, I., Leija-Hernández, G., & López-Bonilla, J., (2014). Rotations in Minkowski space-time. *Bull. Soc. Math. Serv. and Stand, 3*(3), 17–27.

27. Hamilton, W. R., (1844). On quaternions or on a new system of imaginaries in algebra. *Phil. Mag., 25*, 489–495.

28. Cayley, A., (1845). On certain results relating to quaternions. *Phil. Mag., 26*, 141–145.

29. Klein, F., & Sommerfeld, A., (1910). *Über Die Theorie Des Kreisels*. Teubner, Leipzig, [Johnson Reprint Co., New York (1965)].

30. Horn, B. K. P., (1991). Relative orientation revisited. *J. Optical Soc. Am. A, 8*, 1630–1638.

31. Hoggar, S. G., (1992). *Mathematics for Computer Graphics*. Cambridge University Press.

32. Ryder, L. H., (1995). *Quantum Field Theory*. Cambridge University Press.

33. Vince, J., (2011). *Rotations Transforms for Computer Graphics*. Springer-Verlag, London.

34. López-Bonilla, J., López-Vázquez, R., & Prajapati, J. C., (2015). Rotations in three dimensions. *J. of Interdisciplinary Maths., 18*(1/2), 97–102.

CHAPTER 11

Multiscale Approach Towards the Modeling and Simulation of Carbon Nanotubes Networks

MANAS ROY and MITALI SAHA

Department of Chemistry, National Institute of Technology, Agartala – 799046, Tripura, India, Tel.: +91-8974006400, E-mail: mitalichem71@gmail.com (M. Saha)

ABSTRACT

In spite of enormous industrial applications of carbon nanostructures, the greater understanding and control of their synthesis needs to be improved based on the knowledge of carbon-carbon interactions and the interaction of carbon atoms with the other materials. The current production rate remains low, hovering in 10–20%, so an effective process of planning and design for nanomanufacturing is considered necessary for quality assurance of carbon nanostructures and consequentially the yield rate. Among the current technologies, thin films of carbon nanotubes (CNTs) represent one of the most interesting materials from an application-oriented point of view. Therefore, fundamental studies based on theoretical and numerical approaches becomes crucial for the design and development of high-performance devices. This chapter is focused on the basics of different carbon nanotubes (CNTs) technologies from both theoretical and an experimental point of view, for helping in the design of CNT film devices for various applications. This chapter is focused on the theoretical analysis of the networks, based on different models and levels of approximations. All the implemented theoretical models are discussed in order to explain the multiscale approach.

11.1 INTRODUCTION

One of the chemical brilliance of carbon is that it can form diverse types of structures with completely different properties. In our school days, we have learnt about two distinctive allotropes of carbon known diamond and graphite. Eventually, the astonishing breakthrough which modernizes the new area of carbon chemistry was discovering the newer allotrope of carbon which is known as "Fullerenes" (mainly C_{60}) [1]. The discovery of C_{60} was performed by scientist Kroto and Smalley while they made an experiment to investigate the formation of a carbon cluster by the laser-vaporization of graphite. Fullerenes are the closed hollow sphere of the cluster of carbon having pentagonal and hexagonal rings of carbon. 1991 scientist Sumio Iijima observed long, hollow, cylindrical carbon network in Transmission Electron Microscopic (TEM) of the carbon soot obtained by arc-discharge technique while he repeating the experiment proposed by Kratschmer and Huffman for the gram-scale production of "fullerenes" [2]. The long cylindrical hollow helical carbon fiber termed as "carbon nanotubes (CNTs) and considered one of the most enlightening discoveries in scientific community. CNTs are considered as rolled-up helical arrangement of two-dimensional hexagonal graphene sheets and stay markedly dissimilar from conventional carbon fibers.

On the basis of the presence of a number of rolled-up graphene sheet CNTs are classified into two categories one is multi-walled CNTs (MWCNTs) and another is known as single-walled CNTs (SWCNTs). The typical diameter of the cylindrical tube of a SWCNT is about ~1–2 nm whereas MWCNTs are involve with plentiful concentric rolled-up stacks of graphene sheet with outer diameter ~5–50 nm along with interlayer distance of ~ 0.34 nm [3–6]. Both SWCNTs and MWCNTs are few nanometers to several micrometers in length. Both the MWCNTs and SWCNTs have sovereign classifications. MWCNTs are classified as double-walled CNTs, triple-walled CNTs on the basis of the presence of number of inner tubes and SWCNTs are classified as an armchair, zigzag, and chiral on the basis of probable ways of rolling-up of single graphene sheet [3, 6, 7].

CNTs have attained extensive study in enormous purpose owing to their superior structural property and the delocalization of π electron cloud within the two-dimensional graphene sheet produce much stronger C-C bond to impart amazing mechanical strength, electrical as well thermal conductivities. On the other hand, high surface area and presence of mesoporosity within it and its tubular shape make it suitable adsorbent materials for water purification system and CNTs based on bio-medical technology. CNTs have much more mechanical strength than steel but it also much lighter than steel and considered one of the strongest materials. However, there is limited application of CNTs as tensile materials as there are weak van der Waals forces among the inner walls of the CNT along with tubes in CNT bundles. As a result CNTs bundles sliding the CNT walls and reduces the buckling, mechanical, and stiffness property of CNTs. One of the salient features of CNT is that its efficiency depends on mechanical loading along with a suitable perceptive of the mechanics of CNTs is essential to make nanodomain devices and materials.

As the property of atoms and its nanoscopic range are differing so the understanding of the mechanical and electrical property of CNTs on the atomistic point of view is one of the major concern in scientific community. Recently computer modeling like (i) density functional theory (DFT), (ii) molecular dynamic (MD) simulation approach, etc., has been employed to understand nanoscopic property of CNTs. In DFT approximation is done on the basis of ground-state (GS) electronic energy which is distinctive function of the electronic density whereas in MD approximation atoms are considered as classical body having both position and momentum.

11.2 ROLLED-UP APPROACH TOWARDS CONSTRUCTION OF SINGLE-WALLED CARBON NANOTUBES

Rolling-up followed by folding of a single layer graphite sheet (known as graphene) it into the numerous cylindrical structures produces SWCNTs. The rolling-up of the sheets of graphene into SWCNTs has limitation to explain the structural intricacy of CNTs fully. Theoretically, rolling-up of graphene sheet can be performing in numerous ways. One of the essential conditions to make cylinder form graphene strip when Brava is lattice vectors of graphene lattice point (x_1, x_2, x_1, and x_2 are integer) ought to coincide with the origin of the sheet (0, 0). graphene lattice vector is constructed by using following basis vectors:

$$\vec{x_1} = \frac{a}{2}\left(\sqrt{3}, 1\right)$$

(1)

$$\vec{x_2} = \frac{a}{2}\left(\sqrt{3}, -1\right)$$

(2)

where, $a = \sqrt{3}a_{C-C}$ and $a_{C-C} = 1.42 \text{ Å}$ (Carbon-Carbon bond distance in graphite). So, for the rolling-up of graphene sheet two carbon atoms are selected. One of the carbon atoms can be regarded as the origin and then it is thought sheet is rolled up until the two carbon atoms overlap to each other. One can classify the chiral vector ($\vec{c}$, by means of Bravais lattice vectors or basis vector $\vec{x_1}$ and $\vec{x_2}$ in the following approach [8–10].

$$\vec{c} = m_1\vec{x_1} + m_2\vec{x_2} = \left(m_1, m_2\right)$$

(3)

where, m_1 and m_2 are the known as chiral index of the chiral vector.

On the basis of chiral vector and chiral angle (θ), the angle between the chiral vector (c) and basis vector x_1] SWCNTs can be classified into three categories:

i   **Zig-zag CNTs:** For SWCNTs, if the two dimensional graphene sheets has been rolled along anyone symmetry axis of the graphene (say x_1 vector) then the chiral vector express as one of the two basis vectors as (m_1, 0). The (m_1, 0) CNTs termed as Zig-zag CNTs. The chiral angle (θ) for zig-zag CNTs = 0.

ii. **Armchair CNTs:** If the circumferential vector of rolled-up sheet of graphene is situated along the direction between the two basis vectors, i.e., $m_1 = m_2 = m$, then it produces armchair CNTs. In an armchair CNT (m, m) the single graphene sheet is turn round along the chiral vector at an angle 30°.

iii. **Chiral CNTs:** It is the generic nomenclature for SWCNTs which cannot be classified as zig-zag CNTs or armchair CNTs. The chiral angle (θ) of chiral CNTs have any value other than 0° and 30°, i.e., ($\theta \neq 0^{\circ} \neq 30^{\circ}$). In chiral CNT $m_1 \neq m_2 \neq 0$ and it can be express as (m_1, m_2) CNT.

The radius and diameter of any SWCNT can be achieve through the simple geometrical calculation as following:

$$d_{CNT} = \frac{\vec{c}}{\pi} = \frac{\sqrt{3}}{\pi}a_{C-C}\sqrt{m_1^2 + m_1m_2 + m_2^2}$$

(4)

$$T_{CNT} = \frac{\overline{c}}{2\pi} = \frac{\sqrt{3}}{2\pi} a_{C-C} \sqrt{m_1^2 + m_1 m_2 + m_2^2} \qquad (5)$$

where, d_{CNT} and r_{CNT} radius and diameter of SWCNTs.

Thus diameter of armchair CNTs (d_{A-CNT}) and radius armchair CNTs (r_{A-CNT}) of express as:

$$d_{A-CNT} = \frac{\overline{c}}{\pi} = \frac{\sqrt{3}}{\pi} a_{C-C} \sqrt{3m^2} \qquad (6)$$

(As $m_1 = m_2 = m$ in airchair CNT).

$$and \quad d_{A-CNT} = \frac{\overline{c}}{\pi} = \frac{3m}{\pi} a_{C-C} \qquad (7)$$

$$r_{A-CNT} = \frac{\overline{c}}{\pi} = \frac{3m}{2\pi} a_{C-C} \qquad (8)$$

And diameter of Zig-zag CNTs (d_{z-CNT}) and radius armchair CNTs (r_{z-CNT}) of express as:

$$d_{z-CNT} = \frac{\overline{c}}{\pi} = \frac{\sqrt{3}}{\pi} m\, a_{C-C} \qquad (9)$$

$$r_{z-CNT} = \frac{\overline{c}}{\pi} = \frac{\sqrt{3}}{2\pi} m\, a_{C-C} \qquad (10)$$

The relation between chiral angle (θ) with chiral index (m_1, m_2) is given by the following equation (Figure 11.1):

$$\tan(\theta) = \frac{\sqrt{3}m_2}{2m_1 + m_2} \qquad (11)$$

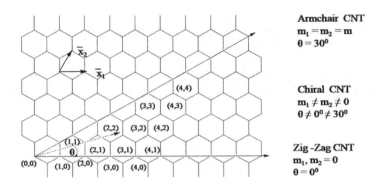

FIGURE 11.1   Rolling-up of single-layer graphene sheet to create different types of SWCNTs [3, 6, 7, 11–15].

11.3   MULTISCALE MODELING APPROACH TOWARDS MULTI-WALLED CARBON NANOTUBES

The rising of technology computer-aided design (TCAD) are mainly devoted to the advancement of fabrication of devices. The importance of TCAD products and its profitable market lies in the fact that it predicts the possibility of the working behavior of a device which is required for the mass production of electronic circuits and the components of any kind of devices. TCAD design has found a special ground in carbon-based materials, especially in the study of CNTs and graphene. Although there are numerous simulation studies for single-walled CNT-based devices, few reports are available on CNT networks. Semi-empirical percolation theory describes the electrical conductivity of a CNT network, according to which the conductivity of a random network is highly non-linear and it has a strong dependence on the density of the nanotubes [16]. However, when the percolation threshold is higher than the nanotubes, the CNTs networks show an insulating behavior due to absence of direct paths between the two electrodes. But on reaching the percolation threshold, the current can flow between the two contacts and the conductivity can be measured up to certain extent.

$$\sigma = \sigma_0 \, (v - v \, c)^t \qquad\qquad (12)$$

where, σ is the electrical conductivity, σ_0 a scaling factor, v the volume fraction of the material, v c the percolation threshold and t an empirical exponent and it depends on the dimensionality of the system. It is noteworthy to mention that some theoretical studies have been done to estimate the percolation threshold values [17, 18] but the other parameters of the equation has not been studied in detail. Besides, this theory lacks of a three-dimensional model for capturing the real morphology of these networks and addresses the homogeneous and infinite networks only. The crucial problem is definitely the generation of a realistic morphology. Although, a two-dimensional model can overestimate the number of percolation paths using many fictitious junctions but the 2-D network can lead to the current flow, different from the real behavior. In case of 3-D model, more reality is observed between the number of the junctions between different CNTs, and therefore the current flow distribution in the network is more reliable. Looking to this aspect, multiscale modeling is required and a detailed study of CNTs would requires quantum mechanical treatment of electron transport and an atomistic description of the nanotube.

The first important step towards a typical simulation workflow is mainly the representation of the device based on the structure. Various approaches can be utilized depending on the problem, for the discretization of the geometry such as stick-percolation theory involving the 3D structure of CNTs. The network generation is mainly done by two consecutive processes, extraction, and the placement of the CNTs. The user may mention the dimensions of device, properties of the solution and the parameters of the deposition, analogs to the experimental setup. After the extraction of lengths, the next step is the extraction of the radii of the CNTs. The last step of generation is setting the number of CNTs to be generated. Before starting the simulation, the program requires to recognize the device as well as kind of simulation to be carried out. However, the program follow a standard routine to convert the morphology file to a proper net list. Before starting any kind of simulation, the device which is going to be simulated and the kind of simulation, it needs, require to be identified. However, the program can follow a routine process to convert the morphology file to a proper net list, independent of the type of simulation, selected. The non-equilibrium greens function (NEGF) formalism is capable to provide computational and conceptual framework to analyze the quantum transport, based on the Schrödinger equation under non-equilibrium conditions. In the past few years, this approach was regularly used to calculate currents and charge densities, both in the coherent and incoherent regime [19–22]. The first step consists in the identification of a suitable Hamiltonian matrix and the second step is to compute the self-energy matrices, and the next step is to compute the retarded Green's function. The final step is to calculate the physical quantities of interest from the Green's function. It is worth to mention that the Hamiltonian matrix is the number of carbon atoms present in CNTs in the device and simulating an entire carbon nanotube computationally, is highly demanding. A tight-binding approximation was used to explain the interaction between the carbon atoms, where only nearest neighbor coupling is considered [19–21].

A further step is the computation of the self-energy matrices for the two contacts after the specification of the Hamiltonian of the device, based on electron-electron and electron-phonon interactions. This concept is also applicable to study the effect of a semi-infinite contact on a device by an opportune modification of the Hamiltonian [22]. The self-energy method is computationally intensive, but it is favorable to treat the open boundary conditions properly. These matrices are energy-dependent and not Hermitian, so are different from the canonical Hamiltonian. After knowing the device Hamiltonian and self-energy matrices, two ways start simulating the system, making use of both the coherent and incoherent transport approximation. The propagation of electrons in a device said to be coherent if it does not face the phase-breaking scattering events. In that case, the effect of the event can be incorporated with an appropriate potential in the Hamiltonian of the system.

11.4   ELECTRICAL AND ELECTROCHEMICAL PROPERTIES OF CNTS

In CNTs the four valence orbital (say $2s$, $2p_x$, $2p_y$ and $2p_z$ and considering z-axis the principle axis) of the C atom are available in bonding. Among these orbital's $2s$, $2p_x$, $2p_y$ combine to give sp^2 hybrid orbital's. These three orbital's are unite themselves to form bonding (σ) and anti-bonding (σ^*) orbital's. The bonding (σ) orbital's are even in regard to planner symmetry as in graphene. The unhybrid $2p_z$ orbital which is considered as odd is unable to form σ orbital. Rather it can undergo lateral interaction with adjacent $2p_z$ orbital construct delocalized bonding (π) and anti-bonding orbital (π^*). The σ bonding orbital in the CNT form hexagonal network in which the carbon atoms are present in the roll-up wall of the cylinder whereas The delocalized π bonds are lie at a 90 angle to the surface of the CNT and are accountable for the very weak interaction of between SWNTs, like two connecting sheets of graphite. Since σ bond are far away from the Fermi energy (E_f) but π^*bands traverse the E_f point at high symmetry corners in the first Brillouin zone and present unsatisfied dangling bonds in graphene sheet [23–25]. Energy accumulates in the unsatisfied dangling bonds of graphene sheet has been decreases during the formation of SWCNTs. The confinement of the delocalized π electron cloud during the rolling-up process may be considering the driving force to create the unique electrical property of SWCNTs. Nevertheless, the electrical property of SWCNTs mainly depends on chirality along with the diameter of the tube. On the basis of condition and bang gap SWCNTs may be metallic or semi-metallic semiconducting and insulating. The electrical conductivity of metallic/semi-metallic SWCNTs are quite very high as there is formation of few surface defects during the rolling-up of graphene sheet and mainly depends on chirality of the rolled-up tubes.

The electronic arrangement along with energy state of graphene sheet can be illustrate by using tight-binding Hamiltonian equation and Zone-folding approximation. The simple mathematical expression for the energy graphene sheet $E_{Graphene}{}^{\pm}(k)$ band-structure can be expressed as [24].

$$E_{Graphene}{}^{\pm}(k)=\pm\gamma_0\sqrt{3+2\cos(k\cdot x_1)+2\cos(k\cdot x_2)+2\cos[k\cdot(x_1-x_2)]} \quad (13)$$

where, γ_0 is called transfer integral between first-neighbors π orbital's having a characteristic value 2.9 eV. x_1 and x_2 known as basis vector express as in Eqns. (1) and (2). k is known as electron momentum. As the quantum electronic state of graphene is entirely 2D sheet having constantly change-able moment in the plane k_x and k_y with distinct acceptable value for k_z then the energy expression will be [24].

$$E_{Graphene}{}^{\pm}(k_x k_y)=\pm\gamma_0\sqrt{1+4\cos(\frac{\sqrt{3}}{2}k_x a)\cos\left(\frac{k_y a}{2}\right)+4\cos^2\left(\frac{k_y a}{2}\right)} \quad (14)$$

where, $a=\sqrt{3}a_{C-C}$ and $a_{C-C}=1.42\,\text{Å}$ (Carbon-Carbon bond distance in graphite) and electron momentum vector in k_x and k_y direction. If we assumed that the valence band (VB) and the conduction band (CB) of graphene sheet congregate at six basis points which are just at the corner of the first Brillouin zone then there is no bang gap between VB and CB and behave like semi-metallic character. For semi-metallic graphene sheet the DOS at the E_f level is non zero. As it has discussed rolled-up form of graphene producing SWCNTs thus the quantized wave vector in the circumferential direction will be:

$$\overline{k}.\overline{c_h} = k_x \cdot c_x + k_y c_y = 2\pi\rho \tag{15}$$

where, ρ is an integer.

This equation provide a relation between k_x and k_y with chiral vector *and* it defines lines in the (k_x, k_y) plane. Each line offers 1D energy band by cutting the 2D graphene sheet in the context of band structure. The particular values of c_x, c_y and ρ find out where these lines traverse the band-structure of graphene sheet. This is most likely the main significant feature of the SWCNTs, which may be also metallic or semiconducting, deciding by whether these imaginary lines go through the E_f points of graphene sheet. It these lines are abandoning to cross the E_f points the SWCNTs, behave like semi-condor, along with a bandgap which can be obtained by the two lines that move toward nearer to the E_f points. Antagonistically, if these lines are intersecting the E_f level, then SWCNTs has crossing bands structure of graphene sheet and are metallic in nature.

The simple mathematical equation we can show metallic or semi-metallic or non-metallic nature of SWCNTs in terms of chirality is given by.

$$|m_1 - m_2| = 3i \tag{16}$$

where, i is an integer number (that is 0.1, 2, 3, 4, ...).

When the indexes (m_1 and m_2) of the chiral vector follow the relation it shows metallic /semi-metallic character.

When $m_1 = m_2$, i.e., i = 0, then VB and CB overlapping with each other to make SWCNTs metallic.

When $m_1 \neq m_2 = 0$ but $|m_1 - m_2| = 3i$ where i ≠ 0 but i = 1, 2, 3.4.........
then SWCNTs is behave like semi-metal.

When $|m_1 - m_2| \neq 3i$ then there must be finite energy gap between VB and CB and band gap close to the E_f level. Then it behaves either semicon-ductor or insulator.

Semi-conducting or insulating SWCNTs largely depends on diameter of the tube (Band gap $\alpha \dfrac{1}{\text{Diameter}}$).

If we consider:

i.   (12, 0)-SWCNTs, i.e., Zig-zag CNT is behaves like semi-metal. As i=4.
ii. (7, 0)-SWCNTs also Zig-zag CNT but it behaves like semiconductor. As I ≠ integer.
iii. (12, 9)-Chiral CNTs it behaves like semi-metal. semi-metal. As i=4.
iv. (12, 8)-Chiral CNTs it behaves like semi-conductor. As I ≠ integer.
v.   (7, 7)-armchair CNTs it behaves metal. semi-metal. As i=0.

The tight binding energy overlap model has limitation in case of very small-diameter SWCNTs. The major cause is associated to the curve of the CNTs, which enhancing by reducing the diameter. Enthrallingly, armchair CNTs always exhibit metallic character even though small diameter. Nevertheless, it has been established that rolled-up the electronic structure of graphene sheet to explain the band structure of SWCNTs construct precise outcome for tubes with diameters (d_{CNT}) larger than 0.8 nm [26–29].

For any semi-conducting SWCNTs the bang gap (Eg) given by following equation.

$$E_{CNT} = 2\frac{E_\infty a_{CNT}}{d_{CNT}} \qquad (17)$$

where, E_∞ is calculated by tight binding energy overlap model and found to be ~2.7 eV. a_{CNT} is the Carbon-Carbon bond distance and d_{CNT} is the diameter of CNT.

For the semi-conducting SWCNTs E_{CNT} are generally vary from ~0.4 ev to 0.8 eV.

11.5 CONCLUSION

This chapter emphasized on the requirement of the multiscale approach towards the simulation and modeling of different types of CNTs for understanding their properties and limitations. A multiscale simulation tool is needed for treating the effects relevant to the carbon nanotubes and simulations can in principle better elucidate some of the most unclear aspects towards an optimum design.

KEYWORDS

- **carbon nanotubes**
- **conduction band**
- **density functional theory**
- **Fermi energy**
- **non-equilibrium greens function**
- **technology computer-aided design**

REFERENCES

1. Kroto, H. W., James, R. H., Sean, C. O., Robert, F. C., & Richard, E. S., (1985). C60: Buckminsterfullerene. *Nature, 318*(6042), 162–163.
2. Iijima, S., (1991). Helical microtubules of graphitic carbon. *Nature, 354*(6348), 56.
3. Dresselhaus, G., & Saito, R., (1998). *Physical Properties of Carbon Nanotubes*: World scientific.
4. Dekker, C., (1999). Carbon nanotubes as molecular quantum wires. *Physics Today, 52*, 22–30.
5. Avouris, P. G. D., & Dresselhaus, M. S., (2000). Carbon nanotubes: Synthesis, structure, properties, and applications. *Topics in Applied Physics.*
6. Dresselhaus, M. S., (1997). Future directions in carbon science. *Ann. Rev. of Mater. Sci., 27*(1), 1–34.
7. Dai, H., (2002). Carbon nanotubes: Synthesis, integration, and properties. *Acc. of Che. Res., 35*(12), 1035–1044.
8. Artyukhov, V. I., Evgeni, S. P., & Boris, I. Y., (2014). Why nanotubes grow chiral. *Nature Commun., 5*, 4892.
9. Zhang, S., Lixing, K., Xiao, W., Lianming, T., Liangwei, Y., Zequn, W., Kuo, Q., Shibin, D., Qingwen, L., & Xuedong, B., (2017). Arrays of horizontal carbon nanotubes of controlled chirality grown using designed catalysts. *Nature, 543*(7644), 234.
10. Dass, D., Rakesh, P., & Rakesh, V., (2012). Analytical study of unit cell and molecular structures of single-walled carbon nanotubes. *International Journal of Computational Engineering Research, 2*(5).
11. Odom, T. W., Jin-Lin, H., Philip, K., & Charles, M. L., (1998). Atomic structure and electronic properties of single-walled carbon nanotubes. *Nature, 391*(6662), 62.
12. Dresselhaus, M. S., Dresselhaus, G., & Saito, R., (1995). Physics of carbon nanotubes. *Carbon, 33*(7), 883–891.
13. Galano, A., (2010). Carbon nanotubes: Promising agents against free radicals. *Nanoscale, 2*(3), 373–380.
14. Calvaresi, M., Mildred, Q., Petra, R., Francesco, Z., & Maurizio, P., (2013). Rolling up a graphene sheet. *Chem. Phys. Chem., 14*(15), 3447–3453.

15. Tung-Wen, C., & Wen-Kuang, H., (2007). Winding of single-walled carbon nanotube ropes: An effective load transfer. *Appl. Phy. Lett., 90*(12), 123102.
16. Stauffer, D., & Aharony, A., (1994). In: Francis, T., (ed.), *Introduction to Percolation Theory.* London: Taylor & Francis, CRC Press.
17. Pike, G. E., & Seager, C. H., (1974). Percolation and conductivity: A computer study. *I. Phy. rev. B, 10*(4), 1421.
18. Seager, C. H., & Pike, G. E., (1974). Percolation and conductivity: A computer study. *II. Phy. Rev. B, 10*(4), 1435.
19. Venugopal, R., Ren, Z., Supriyo, D., Mark, S. L., & Jovanovic, D., (2002). Simulating quantum transport in nanoscale transistors: Real versus mode-space approaches. *J. Appl. Phy., 92*(7), 3730–3739.
20. Anantram, M. P., Mark, S. L., & Dmitri, E. N., (2008). Modeling of nanoscale devices. *Proceedings of the IEEE, 96*(9), 1511–1550.
21. Slater, J. C., & George, F. K., (1954). Simplified LCAO method for the periodic potential problem. *Phy. Rev., 94*(6), 1498.
22. Datta, S., (2005). *Quantum Transport: Atom to Transistor.* Cambridge University Press.
23. Fischer, J. E., & Johnson, A. T., (1999). Electronic properties of carbon nanotubes. *Current Opinion in Solid State and Materials Science, 4,* 28–33.
24. Jean-Christophe, C., Blasé, X., & Roche, S., (2007). Electronic and transport properties of nanotubes. *Rev. Mod. Phy., 79*(2), 677.
25. Ando, T., (2009). The electronic properties of graphene and carbon nanotubes. *NPG Asia Mater., 1*(1), 17–21.
26. Blase, X., Benedict, L. X., Shirley, E. L., & Louie, S. G., (1994). "Hybridization effects and metallicity in small radius carbon nanotubes. *Phy. Rev. Lett., 72*(12), 1878.
27. Cabria, I., Mintmire, J., & White, C., (2003). Metallic and semiconducting narrow carbon nanotubes. *Phy. Rev. B., 67*(12), 121406.
28. Zólyomi, V., & Kürti, J., (2004). First-principles calculations for the electronic band structures of small diameter single-wall carbon nanotubes. *Phy. Rev. B, 70*(8), 085403.
29. Connétable, D., Rignanese, G. M., Charlier, J. C., & Blasé, X., (2005). Room temperature peierls distortion in small diameter nanotubes. *Physical Review Letters, 94*(1), 015503.

CHAPTER 12

When Perovskites Memorize

P. SIDHARTH, M. PRATEEK, and P. PREDEEP

*Laboratory for Molecular Electronics and Photonics (LAMP),
Department of Physics, National Institute of Technology,
Calicut, Kerala, India, E-mail: predeep@nitc.ac.in (P. Predeep)*

ABSTRACT

Intense researches on perovskite solar cells have achieved an impressive efficiency of 23.3%. Apart from solar cells, perovskite materials are finding more and more applications like that in LED, FET, Nano LASER, Photodetectors, etc. Recently, hectic investigations are reported on the memory and switching potential of perovskites. Materials showing biostability by resistive switching are the ideal candidates for memory device fabrication. Perovskite material shows J-V hysteresis, have a tuneable bandgap and long electron-hole diffusion length. Along with these properties and low cost, synthesis methods make them promising candidates for next-generation volatile and non-volatile memory devices. In this chapter, the characteristic aspects of memory devices, in general, are first described and the same has been extended to the memory effects in perovskites. This is followed by a discussion on different types of conduction mechanisms in perovskite materials under a bias voltage which causes resistive switching. To analyze the advancement in perovskite memory devices, different perovskite memory devices fabricated so far are considered. Their synthesis methods, different layers used for electron-hole transport, memory properties like on/off ratio, retention time, operating voltage, and power consumption are critically evaluated. The future scope of perovskite materials in memory device fabrication is also discussed.

12.1   INTRODUCTION

Perovskite materials received rapid attention and generated great interest with solar cells made with them as absorbers crossed and an impressive 20% efficiency mark [1]. Since then newer application possibilities for these versatile materials are being explored, like that in LEDs [2], Detectors [3, 4], switchable memories [5], etc. Out of these, memory application of Perovskites seems to be the most promising. Though in a fledgling state with much to go before establishing as a technologically and commercially viable memory alternative, Perovskites memory research, as in the case of Perovskites solar cells had made large strides in a short period and has become a fast-emerging field of research. This chapter is intended to provide a brief foray into these emerging trends in the exploration of memory application of perovskite materials.

Resistive Random Access Memory (RRAM) is considered as the most promising next-generation nonvolatile memory because [6, 7] of their low operating voltage, fast switching speed, high on/off ratio, low power consumption, complementary metal-oxide-semiconductor (CMOS) compatibility and simple sandwich architecture. Perovskite materials with their high absorption coefficient, tunable bandgap, high electron-hole diffusion length, ambipolar charge transport, and low-cost synthesis methods promises [8, 9] to be ideal materials for RRAMs. Intense research is going on in organic-inorganic halide perovskites and all-inorganic perovskites for memory application. Results, so far reported are more than impressive and making larges strides forward from simple memory devices to high-density multi-level storage memory devices and artificial synapses.

The hysteresis, often found in the J-V characteristics of perovskite solar cells is considered as an important performance obstacle. However, ironically the same characteristics carry the signature of memory effects in perovskites. Space charge traps that are a burden in solar cells which causes J-V hysteresis is the sign of memory behavior, and perovskite materials used in solar PV have plenty of them and their mitigation is considered to be one of the major challenges with PV application. However such unfavorable characteristics of perovskite are making them favorable in applications like memory switching and this shows the high degree of versatility possessed by the perovskites.

Before going to the details of perovskite memory devices, a brief overview of the general memory principles will not be out of place here. The chapter therefore is planned as follows: The discussions begin with brief ideas about basic memory device structures, operating principles, conduction mechanism, and details of mechanisms behind resistive switching. This will be followed by a brief outlook about perovskite materials that are widely in use in memory research so far. Then the state of the art in memory explorations with perovskites is analyzed by comparing different perovskite memory devices reported under three basic categories, viz., mixed halide perovskites, Trihalide perovskites, and all-inorganic perovskites. In the final part different characteristics of memory devices such as on/off ratio, operating voltage, retention time, and endurance are compared from reported devices. Brief notes on the future and further scope of the area conclude the chapter.

12.2 CHARACTERISTIC ASPECTS OF A GENERAL MEMORY DEVICE

12.2.1 STRUCTURE OF MEMORY DEVICES

The general structure of memory devices employs a sandwich structure where an active material is sandwiched between two electrodes. The bottom electrode is usually Indium Tin oxide (ITO) or Fluorine doped FTO, and top electrode is usually a metal. Memories can be classified into organic memory, inorganic memory, polymer memory, hybrid organic-inorganic perovskite memory, inorganic perovskite memory, etc., based on the active material used in the device. Apart from this basic structure, a lot of devices have been reported incorporating different anode and cathode buffer layers to enhance memory performance. Figure 12.1 shows [10] structural and energy level diagram of an organic memory in which the active material is a combination of $F_{16}CuPc$, PbPc, Bphen, and CuI as anode and cathode buffer layers respectively.

As already mentioned, Resistive switching is the underlying mechanism in these new generation unconventional memory devices, and we could identify the underlying mechanism behind the resistive switching behavior of the device employing theoretical models like space charge limited conduction (SCLC), Schottky thermionic emission, or Fowler Nordheim tunneling through proper fitting process of the corresponding J-V curves.

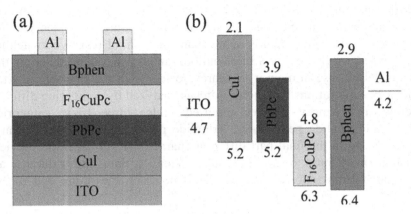

FIGURE 12.1   (a) Structure and (b) schematic energy level diagram of Device.
Source: Reproduced with permission [10]. Copyright 2017, Royal Society of Chemistry.

12.2.2   RESISTIVE SWITCHING

Unlike the conventional silicon-based memory devices, where data get
stored in the form of stored charges, in organic and perovskite polymer
memories data are stored as different resistance states. Usually, these
materials have two resistance states namely a high resistance state (HRS)
and a low resistance state (LRS). The shifting from one resistance state
to other is known as resistive switching. Without any bias voltage, the
device is in the HRS state. On applying a bias voltage, device stays in
HRS for low voltages and shift to LRS at some threshold voltage, also
called the SET voltage. On reversing the bias voltage, devices behave in
two ways; one, where the device stays in LRS irrespective of increasing
reverse voltage, which is a volatile memory, also called Write Once Read
Many (WORM) devices [11], and the other, where device can come back
to HRS as reverse voltage is increased which gives a non-volatile memory.
Different mechanisms have been proposed to explain the conduction
mechanism in these materials, mainly space charge traps and filamentary
conduction. One can explore the underlying mechanisms behind resistive
switching and extract more physical quantities involved in the process by
appropriately fitting the experimental data with theoretical models like
SCLC, Schottky emission, Fowler-Nordheim tunneling, etc. SCLC is the
most reported theoretical model behind resistive switching in the case of

perovskite memory devices and filament formation is the most proposed conduction mechanism.

12.2.2.1 SPACE CHARGE TRAPS

When the contact between a metal electrode and a trap free active material is ohmic, injected charge carriers from the electrode will accumulate near the electrode-active layer interface. The mutual repulsion between these charge carriers can restrict further injection of charge carriers from electrodes. So a space charge channel is built in the active material and the conduction is controlled by these space charges [12]. When the bias is applied to a device in HRS, conduction is mainly due to the intrinsic thermally generated charge carriers. As the bias voltage increases the traps in the active layers starts to fill by the injected charge carriers. Once all the traps are filled, a space charge channel will be built along the active material with injected charge carriers being the conductors, thus shifting the device to LRS. On applying a reverse voltage, if the space charge channel gets removed and device goes back to HRS, it will form a non-volatile memory, and if the space charge channel is not broken and device stays in LRS will be a volatile WORM memory. Space charges in materials are present due to many reasons like injection from electrodes, accumulation of ions at the interface, ionized do pants in interfacial regions, etc. Traps may also be present in active material and at interfaces, and they can reduce charge carrier mobility and at interfaces, they can affect [13] the charge injection from electrodes.

12.2.2.2 FILAMENTARY CONDUCTION

Resistive switching by filamentary conduction is reported in two different ways: Metal filament formation in the case of devices using a silver [14, 15] or aluminum [16, 17] electrode, and filament formation due to [7, 18, 19] the migration of halide vacancies in halide-based organic-inorganic hybrid perovskites. Conduction mechanism is almost the same in both cases. In devices with silver electrodes, on applying a bias voltage, oxidized Ag ions from the electrode are injected into the bulk material, and they move across the device towards the counter electrode, and get deoxidized by combining with either injected electrons or thermally generated intrinsic electrons thus forming a conductive filament. Current flows through these filaments and LRS is obtained. On applying reverse voltage rupturing of the filaments, begin to happen as Ag ion moves back to the electrode dioxides into Ag atoms. The device goes back to HRS when filaments are removed [6. 14, 15].

In halide based organic-inorganic perovskites, when an electric field is applied, halide ions accumulate near anode, and randomly distributed halide vacancies arrange themselves across the material in the form of filaments, thus forming the conduction filament. Conduction being through the filaments, it shifts the state to LRS. On the application of reverse field, halide ions moves back to bulk and fill the vacancies thus rupturing the conductive filaments and the device goes back [5, 20, 21] to HRS.

12.2.3   CONDUCTION MECHANISMS

To understand more about the underlying mechanisms and extract physical quantities related to them, fitting of the I-V curve with different theoretical models is done. Various types of conduction and fitting proceedings are briefed in this section.

12.2.3.1   OHMIC CONDUCTION

For low bias voltages, weak injection from electrodes couldn't fill all the traps, so ohmic conduction is dominant. Here I–V curve shows a linear behavior.

12.2.3.2   TRAP-FILLED LIMIT REGION (TFL)

TFL is usually observed right after the ohmic region. There is a sudden increase in current with bias voltage, the injected electrons begin to fill the traps along with the rising voltage until the voltage is large enough to fill all the traps, and the voltage is then called the VTFL (trap-filled limited voltage). This process is also named the "set process" and the device transfers from HRS to LRS accordingly. Here conduction obeys Mark Helfrich equation [22]:

$$J = q^{1-l} \mu N_C \left(\frac{2l+1}{l+1} \right)^{l+1} \left(\frac{l}{l+1} \frac{\varepsilon \varepsilon_0}{H_t} \right)^l \frac{V^{l+1}}{d^{2l+1}}$$

where; q is the electronic charge, $l = \frac{T_C}{T}$ where T_C is the characteristic temperature and T is the measuring temperature, μ is the carrier mobility, N_C is the density of states in the relevant band, ε_0 is the vacuum permittivity, ε is the dielectric constant, H_t is the trap density, V is the applied bias voltage and d is the device thickness.

12.2.3.3 SPACE CHARGE LIMITED CONDUCTION (SCLC)

When the injected electrons have filled all the traps, SCLC conduction is dominant. Here, the injected charge carrier density exceeds the intrinsic charge carrier density of the material [23]. In this region, voltage-current relationship obeys Child's law; $I \propto V^2$. Conduction in SCLC region can be described using Motte-Gurney law [24, 25]:

$$J = \frac{9}{8} \varepsilon_0 \varepsilon \mu \frac{V^2}{d^3}$$

where, J is the current density, ε_0 is the vacuum permittivity, ε is the dielectric constant, μ is the electronic mobility, V is the applied bias voltage and d is the device thickness.

12.2.3.4 SCHOTTKY THERMIONIC EMISSION

When a metal-semiconductor contact is created, it will lead to an energy barrier at the contact. If the applied electric field is large enough to give energy to electrons to jump over the barrier then we can say there is a field-assisted thermionic injection. This is Schottky emission. This type of charge injection is preferred at high temperatures or at low barrier heights. Schottky emission current-voltage dependency is given as [26]:

$$log I \propto \sqrt{V}$$

12.2.3.5   FOWLER-NORDHEIM TUNNELING

Along with Schottky emotion at metal-insulator contacts, electrons can tunnel through the energy barrier if the applied field is large enough and the barrier width is small accordingly. This is a quantum mechanical phenomenon. This type of tunneling is called Fowler Nordheim tunneling. The current density voltage dependence is given as [26, 16]:

$$log\left(\frac{J}{V^2}\right) \propto \frac{1}{V}$$

12.2.4   MEMORY PARAMETERS

- **On/Off Ratio:** It is one of the most important parameters when it comes to a good memory device. It is the ratio of the current in LRS to current in HRS. It can be measured from a simple endurance curve (current vs. number of cycles). A high and stable On/Off ratio is the most desired characteristic of a practical memory device. The more the On/Off ratio the more the difference between SET current and RESET current which ensures a good memory performance as the reading of SET, RESET voltages will be faster, error-free and less power consuming [27]. Improving the value of On/Off ratio is one of the major challenges in perovskite memory devices. A large hysteresis is ideal to get a good On/Off ratio. Xu et al. [6] reported that materials having large trap density show a larger hysteresis.

- **Operating Voltage:** It is the voltage at which device shift from HRS to LRS. A low operating voltage is expected from an ideal memory device. The lower the operating voltage, lower will be the power consumption of the device. A low trap density is required to have a low operating voltage because, higher the trap density higher will be the voltage required to fill the traps/form conductive filaments.

- **Retention Time:** It is very important when it comes to practical applications. It is the measure of device's capacity to hold the data once written. Improvement of retention time is one of the much wanted aspect area when it comes to perovskite memory devices.

- **Endurance Time:** It tells us about the stability of the device. It is usually measured in terms of the no. of cycles. A large endurance time is expected from a good memory device.

Devices showing excellent results in any one of the above parameters are already reported. However, perovskite memory devices with reasonable values of all the above three mentioned characteristics are yet to be reported. The challenge is developing perovskite memory devices having a large On/Off ratio, retention time, endurance time and a low operating voltage.

In the following section discussion on reported memory devices with perovskites is presented.

12.3 PEROVSKITE MEMORY DEVICES

12.3.1 PEROVSKITE MATERIALS

Before moving on to the different perovskite memory devices and the state of the art, for the comfort and convenience of the readers, let's have a very brief discussion about perovskite materials.

The first perovskite material $CaTiO_3$ was discovered in early 1839 by German mineralogist Gustav Rose and named after the Russian mineralogist Lev Perovski [28]. The general chemical formula of perovskites is [29, 30] ABX_3 where A monovalent large cation, B is a smaller divalent cation and X is the anion, usually halogens or oxides. These perovskites have cubical or tetragonal structure. A and B cations are octahedrally coordinated with X anions to form the basic building block. A cation is embedded in the space between nearby BX_6 octahedron connected in a 3D corner-sharing configuration. A site cation fills the gap between the connected octahedra and neutralizes the charge of the structure [28, 30]. Perovskites can be classified into inorganic perovskites where both A and B cations are inorganic, e.g., $CsPbBr_3$ (Figure 12.2a) and organic-inorganic hybrid perovskites where A cation is an organic cation, e.g., $CH_3NH_3PbX_3$ (Figure 12.2b) [30] Perovskites show strong optical absorption, long-range ambipolar charge transport, long electron-hole diffusion length, low excitation binding energy, tunable bandgap, high dielectric constant, ferroelectric properties, etc.

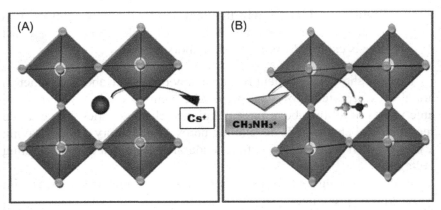

FIGURE 12.2   (a) Inorganic Perovskite structure, (b) Organic-inorganic perovskite structure.
Source: Reproduced with permission [30]. Copyright 2018, Elsevier.

A variety of perovskite structures showing resistive switching have been widely investigated for memory applications. Among them, transition metal oxide perovskites and organic-inorganic halide perovskites are of high interest. Transition metal oxide perovskites ABO_3 such as $BaTiO_3$, $SrTiO_3$, $BiFeO_3$, $SrZrO_3$, and $Pr_xCa_{1-x}MnO_3$ showed [8, 19] very good resistive switching effects and stability yet their high-temperature preparations and fragile nature limit their application as memory devices. Organic-inorganic halide perovskites can be prepared using simple solution processing techniques and comparatively low annealing temperatures (not higher than 150°C) that make them a popular choice.

In the following sections, different perovskite memories are discussed by classifying them into three categories, Mixed halide perovskite memory devices; in which the X anion is a mixture different halide, Trihalide perovskite memory devices; in which X anion is a single halide and All inorganic perovskite memory devices; in which both A and B cations are inorganic ions and X is a halide ion.

12.3.2   *MIXED HALIDE PEROVSKITE MEMORY DEVICES*

In 2015, Yoo et al. [5] reported the first resistive switching memory using organic-inorganic hybrid perovskites. They used a chlorine doped methylammonium (MA) lead iodide perovskite to fabricate a memory cell with FTO substrate and Au as the top electrode. This device with the structure, $Au/CH_3NH_3PbI_{3-x}Cl_x/FTO$ was operated at a bias voltage of 0.8 V and showed great retention characteristic of 10^4 seconds and an endurance of 100 cycles. They also studied [5] the temperature dependence of memory characteristics and found that the device is stable up to 80°C even though, the on-off ratio was less than 1 which is too poor.

Shaban et al. [18] studied the switching mechanism on a chlorine doped MA lead iodide device using different probes of silver epoxy, copper, and graphite (Figure 12.3). All the probes showed resistive switching but only the silver probe showed consistent memory behavior all over the perovskite layer.

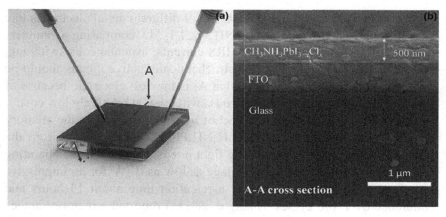

FIGURE 12.3 (a) Test setup used for the characterization of the probe-based resistive memory cell. (b) SEM cross-section image of $CH_3NH_3PbI_3-xCl_x$ layer formed on FTO/glass substrate.
Source: Reproduced with permission [18]. Copyright 2019, Elsevier.

They examined the effect of temperature on LRS (Figure 12.4) and concluded that it was iodide ion migration causing filament formation that causes resistive switching. They obtained a high on-off ratio of 10^6 in the first cycle, which then decreased to less than 100 in subsequent cycles.

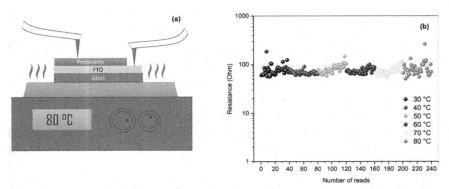

FIGURE 12.4 (a) Test structure used to investigate LRS temperature dependence. (b) Resistance of the LRS in $CH_3NH_3PbI_3-xClx/FTO$ structure under different temperatures. *Source*: Reproduced with permission [18]. Copyright 2019, Elsevier.

Yan et al. [27] investigated the effect of different metal electrodes like Au, Ag, Zn, Cu, Ti, and Al on $CH_3NH_3PbCl_xI_{3-x}$ by comparing parameters like SET, RESET voltages, LRS, HRS currents, instantaneous switching, and the costs. They stated that electrochemically active metals should be used as electrodes and concluded that Al is the best electrode because of low operational voltage, instantaneous switching, at relatively low cost.

Zhou et al. [19] studied the effect of light illumination on the memory device. They illuminated an $Au/CH_3NH_3PbI_{3-x}Cl_x/FTO$ device from the FTO side and found that increasing light intensity can decrease operating voltage. They got an operating voltage as low as 0.1 V for an impressive on-off ratio of 10^4 which showed a retention time about 13 hours and endurance over 400 cycles. Huang et al. [14] reported a memory device of the structure $Ag/FA_{.83}MA_{.17}Pb(I_{.82}Br_{.18})_3/FTO$, which could be used in artificial synapses. They also showed that the mechanism behind switching behavior is Ag ions migration across the device using an energy-dispersive X-ray spectroscopy technique. Their on-off ratio and endurance were too poor which was less than 10 and 40 cycles, respectively.

12.3.3 TRIHALIDE PEROVSKITE MEMORY DEVICES

MA Lead Triodide ($MAPbI_3$) is the most popular perovskite material. Choi et al. [31] showed that multilevel data storage is possible in a single device by controlling the compliance current. They fabricated a device with structure $Pt/Ti/SiO_2/Si/CH_3NH_3PbI_3/Ag$ operated at a voltage of 0.13 V and showed an on-off ratio of 10^6. Xu et al.[6] studied the effect of defect density and hysteresis on $MAPbI_3$ perovskite memory devices by adding a MoO_3 buffer layer between $MAPbI_3$ and Ag electrode. The fabricated memory cell structure $ITO/MAPbI_3/MoO_3/Ag$, operated at 0.25 V and yielded an on-off ratio of 10^2. It was found that defect density affected operating voltage and the hysteresis behavior were critical for on-off ratio values. However, a flexible device on a PET substrate with almost similar properties except a lesser brought an on-off ratio of only just 15. Ruan et al. [8] doped Bismuth into MA lead bromide and used a device structure of $FTO/ZnO/CH_3NH_3Pb_{1-x}Bi_xBr_3/Pt$ to get an impressive on-off ratio of 10^5 at an operating voltage 1.05 V where the device without ZnO yielded an on-off ratio of maximum 20 at an operating voltage 0.85 V. Further, both the devices showed an endurance of only 100 cycles. Interestingly they reported two types of polarity during the endurance test; anticlockwise polarity in which device switch from HRS to LRS, and clockwise polarity where device switched from LRS to HRS. The device also showed good thermal stability up to 80°C.

Stability in humid atmospheric conditions has always been the challenge faced by perovskite devices. Bohee Hwang and Jang-Sik Lee [32] reported that passivation of ZnO or ALD-AlO_x layers in $MAPbI_3$ increases the device stability up to 30 days in the air. In another study [33] they [34] the same authors demonstrated a lead free perovskite memory device using bismuth which may answer the question about environmental issues arising due to the usage of toxic lead. The device $Au/CH_3NH_3)_3Bi_2I_9/ITO$ also showed [33] multilevel data storage, having three distinct resistance states for different compliance current (Figure 12.5). On-off ratio was just 100 for the device at an operating voltage 1.6 V and endurance was 300 cycles although the device showed a good retention time of 10000 seconds.

12.3.4 ALL INORGANIC PEROVSKITE MEMORIES

Recent developments in perovskite memories are highly focused on all inorganic memory devices. A multilevel memory approach where multiple physical qualities are used for multilevel storage in a single device could be the best candidate for the development of high-density memory devices. Chen et al. [7] developed a device using $CsPbBr_3$ quantum dots with structure $Au/CsPbBr_3$ QDs /ITO, which is capable of electronic controlled multilevel

data storage and light assisted multi-level data storage. This device obtained
an on-off ratio of 107 at an operating voltage-2.4 V and showed a retention
time of 10000s with an endurance over 500 cycles. Liu et al. [19] developed
a flexible all-inorganic perovskite using cesium as A cation.

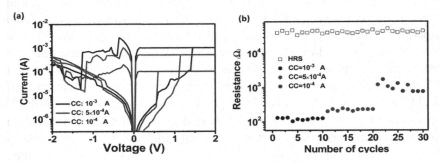

FIGURE 12.5   (a) Typical I–V characteristics of the memory device under different
compliance currents of 1 mA, 500 μA, and 100 μA. (b) Multilevel resistance states with
different compliance currents [33].
Source: Reproduced with permission. Copyright 2018, Royal Society of Chemistry.

The device Al/CsPbBr$_3$/ PEDOT: PSS/ITO/PET [19] showed consistent
memory characteristics while bending at different angles (Figure 12.6).
Also, the device has a clockwise polarity, the set process to occur at-0.6
V, though the on-off ratio, 100 endurance, and 50 cycles were too poor
for this device. Zhuang Xiong et al. [34] reported a lead free inorganic
memory device which showed good stability over two months in ambient
conditions. The Al/CsBi$_3$I$_{10}$/ITO device operated at-1.7 V with an on-off
ratio of 103, retention time 104 and poor endurance over just 150 cycles.

12.3.5   ROLE OF BUFFER LAYERS IN MEMORY PERFORMANCE

In organic memory devices, different anode and cathode buffer layers are
used extensively to enhance memory performance. Perovskite memory
devices are also reported having buffer layers but not as extensively as in
organic memory devices. It can be seen that the incorporation of buffer
layers provided good changes in memory parameters of those devices.

Choi et al. [35] used a Pt/Ti/SiO2/Si/CH3NH3PbI3/Ag device which
yielded a high on-off ratio of 106 but the role of Pt/Ti/SiO2/Si in this
achieving this is not mentioned in their study. Yan et al. [27] reported a

MAPbI3 device with Al and FTO as electrodes. Here it is introduced a buffer layer of TiO2 between FTO and perovskite, which helped in reducing the HRS current and thereby boosting on-off ratio up to 109 which is the best reported yet in perovskite memory devices. TiO2 has been used in solar cells as a function layer to reduce leakage of charges, in the case of memory devices; role of TiO2 is yet to be asserted. ZiqiXu et al. [6] used MoO3I as buffer layer to reduce the energy gap between metal and perovskite active layer which made charge injection easier thereby decreasing the operating voltage. Bohee Hwang and Jang-Sik Lee [32] reported the use of ZnO layer below the top metal electrode as a protective layer and succeeded in getting a higher on-off ratio. However, ZnO doesn't influence the switching behavior. Ruan et al. [8] also used ZnO as a buffer layer this time between bottom FTO electrode and perovskite active layer and found that the on-off ratio increased from 20 to 105 due to the introduction of ZnO buffer layer. It is clear that incorporation of buffer layers increases memory performance. Most of the buffer layers are chosen to decrease energy gap between electrode and active layer, thereby smoothening charge carrier injection, or as layers to protect active layer from degrading. A deeper understanding of the mechanism in these buffer layer-active layer interfaces is still to be explored. A comparison of the performance parameters of different perovskite memory devices is given in Table 12.1.

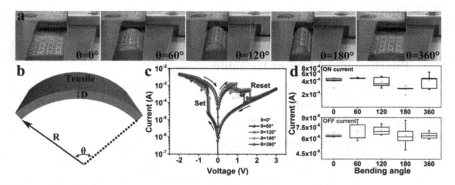

FIGURE 12.6 All-inorganic perovskite CsPbBr3 ReRAM with different bending angles. (a) Photographs of flexible Al/CsPbBr3/PEDOT: PSS/ITO devices being bent. (b) Schematic illustration of the flexible substrate at tensile strain (up) where R is the bending radius, D is the thickness of the devices, and θ is the corresponding central angle. (c) I−V characteristics and (d) bending stability with different bending angles of perovskite resistive switching memory with repetitive bending cycles.

Source: Reproduced with permission [19]. Copyright 2017, American Chemical Society.

It can be seen from the table that FTO electrode and buffer layers play a critical role in getting impressive on-off ratios. Only in one device with ITO the memory registered impressive switching ratio, and in that case, the active layer consisted of QDs. This observation raises an important aspect to be the investigated-the role of ITO and FTO electrodes in memory performance.

12.4   FUTURE SCOPE OF PEROVSKITE MEMORIES

Industry viable on-off ratio and other parameters of perovskite memory devices are still wanting and much scope is in this aspect. Memory devices incorporating buffer layers aren't reported much. This area for further research. Though lead-free perovskites are being reported, this area also needed further research to obtain stable and having good memory characteristics for lead-free devices. Solar cell research based on perovskite materials is exploring the possibilities of single-crystal perovskite wafers for direct application as an absorbing layer. Application of such layers is yet to be explored in perovskite memory devices. This is another interesting aspect that offers lot of scope for further investigations in the topic.

KEYWORDS

- complementary metal-oxide-semiconductor
- fluorine-doped indium tin oxide
- high resistance state
- low resistance state
- resistive random access memory
- space charge limited conduction

REFERENCES

1. Chen, W., Wu, Y., Yue, Y., Liu, J., Zhang, W., Yang, X., Chen, H., et al., (2015). Efficient and stable large-area perovskite solar cells with inorganic charge extraction layers. *Science, 350*(80), 944–948. doi: 10.1126/science.aad1015.

2. Xiao, Z., Kerner, R. A., Zhao, L., Tran, N. L., Lee, K. M., Koh, T. W., Scholes, G. D., & Rand, B. P., (2017). Efficient perovskite light-emitting diodes featuring nanometer-sized crystallites. *Nat. Photonics, 11*, 108–115. doi: 10.1038/nphoton.2016.269.

3. Sutherland, B. R., Johnston, A. K., Ip, A. H., Xu, J., Adinolfi, V., Kanjanaboos, P., & Sargent, E. H., (2015). Sensitive, fast, and stable perovskite photodetectors exploiting interface engineering. *ACS Photonics, 2*, 1117–1123. doi: 10.1021/acsphotonics.5b00164.

4. Wei, H., Fang, Y., Mulligan, P., Chuirazzi, W., Fang, H. H., Wang, C., Ecker, B. R., Gao, Y., Loi, M. A., Cao, L., & Huang, J., (2016). Sensitive X-ray detectors made of methylammonium lead tribromide perovskite single crystals. *Nat. Photonics, 10*, 333–339. doi: 10.1038/nphoton.2016.41.

5. Yoo, E. J., Lyu, M., Yun, J., Kang, C. J., Choi, Y. J., & Wang, L., (2015). Resistive switching behavior in organic-inorganic hybrid $CH_3NH_3PbI_{3-x}Cl_x$ perovskite for resistive random access memory devices. *Adv. Mater., 27*, 6170–6175. doi: 10.1002/adma.201502889.

6. Xu, Z., Liu, Z., Huang, Y., Zheng, G., Chen, Q., & Zhou, H., (2017). To probe the performance of perovskite memory devices: Defects property and hysteresis. *J. Mater. Chem. C., 5*, 5810–5817. doi: 10.1039/c7tc00266a.

7. Chen, Z., Zhang, Y., Yu, Y., Cao, M., Che, Y., Jin, L., Li, Y., et al., (2019). Light assisted multilevel resistive switching memory devices based on all-inorganic perovskite quantum dots. *Appl. Phys. Lett., 114*. doi: 10.1063/1.5087594.

8. Ruan, W., Hu, Y., Xu, F., & Zhang, S., (2019). Resistive switching behavior of organic-metallic halide perovskites $CH_3NH_3Pb1-xBixBr_3$. *Org. Electron. Physics, Mater. Appl., 70*, 252–257. doi: 10.1016/j.orgel.2019.04.031.

9. Gu, C., & Lee, J. S., (2016). Flexible hybrid organic-inorganic perovskite memory. *ACS Nano, 10*, 5413–5418. doi: 10.1021/acsnano.6b01643.

10. Kong, Z., Liu, D., He, J., & Wang, X., (2017). Electrode buffer layers producing high performance nonvolatile organic write-once-read-many-times memory devices. *RSC Advances*, 13171–13176. doi: 10.1039/c7ra00764g.

11. Prime, D., & Paul, S., (2009). Overview of organic memory devices. *Philos. Trans. R. Soc. A Math. Phys. Eng. Sci., 367*, 4141–4157. doi: 10.1098/rsta.2009.0165.

12. Yen, H. J., Shan, C., Wang, L., Xu, P., Zhou, M., & Wang, H. L., (2017). Development of conjugated polymers for memory device applications, Polymers (Basel). 9. doi: 10.3390/polym9010025.

13. Taunk, M., & Chand, S., (2014). Bias and temperature dependent charge transport in flexible polypyrrole devices. *J. Appl. Phys., 115*. doi: 10.1063/1.4866329.

14. Huang, Y., Zhao, Z., Wang, C., Fan, H., Yang, Y., Bian, J., & Wu, H., (2019). Conductive metallic filaments dominate in hybrid perovskite-based memory devices. *Sci. China Mater., 62*, 1323–1331. doi: 10.1007/s40843-019-9433-4.

15. Yoo, E., Lyu, M., Yun, J. H., Kang, C., Choi, Y., & Wang, L., (2016). Bifunctional resistive switching behavior in an organolead halide perovskite based Ag/CH 3 NH 3 PbI_3-x Cl x /FTO structure. *J. Mater. Chem. C, 4*, 7824–7830. doi: 10.1039/c6tc02503j.

16. Prakash, A., Ouyang, J., Lin, J. L., & Yang, Y., (2006). Polymer memory device based on conjugated polymer and gold nanoparticles *J. Appl. Phys., 100*. doi: 10.1063/1.2337252.

17. Baral, J. K., Majumdar, H. S., Laiho, A., Jiang, H., Kauppinen, E. I., Ras, R. H. A., Ruokolainen, J., Ikkala, O., & Österbacka, R., (2008). Organic memory using [6,6]-phenyl-C61 butyric acid methyl ester: Morphology, thickness, and concentration dependence studies. *Nanotechnology, 19*. doi: 10.1088/0957-4484/19/03/035203.

18. Shaban, A., Joodaki, M., Mehregan, S., & Rangelow, I. W., (2019). Probe-induced resistive switching memory based on organic-inorganic lead halide perovskite materials. *Org. Electron. Physics, Mater. Appl., 69*, 106–113. doi: 10.1016/j.orgel.2019.03.019.

19. Liu, D., Lin, Q., Zang, Z., Wang, M., Wangyang, P., Tang, X., Zhou, M., & Hu, W., (2017). Flexible all-inorganic perovskite $CsPbBr_3$ nonvolatile memory device, *ACS Appl. Mater. Interfaces, 9*, 6171–6176. doi: 10.1021/acsami.6b15149.

20. Tian, H., Zhao, L., Wang, X., Yeh, Y. W., Yao, N., Rand, B. P., & Ren, T. L., (2017). Extremely low operating current resistive memory based on exfoliated 2D perovskite single crystals for neuromorphic computing. *ACS Nano, 11*, 12247–12256. doi: 10.1021/acsnano.7b05726.

21. Zhu, X., Lee, J., & Lu, W. D., (2017). Iodine vacancy redistribution in organic-inorganic halide perovskite films and resistive switching effects. *Adv. Mater., 29*, 1–8. doi: 10.1002/adma.201700527.

22. Kumar, V., Jain, S. C., Kapoor, A. K., Poortmans, J., & Mertens, R., (2003). Trap density in conducting organic semiconductors determined from temperature dependence of J-V characteristics. *J. Appl. Phys., 94*, 1283–1285. doi: 10.1063/1.1582552.

23. Joung, D., Chunder, A., Zhai, L., & Khondaker, S. I., (2010). Space charge limited conduction with exponential trap distribution in reduced graphene oxide sheets. *Appl. Phys. Lett., 97*, 1–3. doi: 10.1063/1.3484956.

24. Diaham, S., & Locatelli, M. L., (2012). Space-charge-limited currents in polyimide films. *Appl. Phys. Lett., 101*. doi: 10.1063/1.4771602.

25. Tuladhar, S. M., Poplavskyy, D., Choulis, S. A., Durrant, J. R., Bradley, D. D. C., & Nelson, J., (2005). Ambipolar charge transport in films of methanofullerene and poly(phenylenevinylene)/methanofullerene blends. *Adv. Funct. Mater., 15*, 1171–1182. doi: 10.1002/adfm.200400337.

26. Emtage, P. R., & Tantraporn, W., (1962). Schottky emission through thin insulating films. *Phys. Rev. Lett., 8*, 267–268. doi: 10.1103/PhysRevLett.8.267.

27. Yan, K., Peng, M., Yu, X., Cai, X., Chen, S., Hu, H., Chen, B., Gao, X., Dong, B., & Zou, D., (2016). High-performance perovskite memristor based on methyl ammonium lead halides. *J. Mater. Chem. C, 4*, 1375–1381. doi: 10.1039/c6tc00141f.

28. Watthage, S. C., Song, Z., Phillips, A. B., & Heben, M. J., (2018). Chapter 3 evolution of perovskite solar cells. In: Thomas, S., & Thankappan, A., (eds.), *Perovskite Photovoltaics* (pp. 43–88). doi: 10.1016/b978-0-12-812915-9.00003-4.

29. Varma, P. C. R., (2018). Chapter 7-low-dimensional perovskites. In: Thomas, S., & Thankappan, A., (eds.), *Perovskite Photovoltaics* (pp. 197–229). Academic Press. doi: https://doi.org/10.1016/B978-0-12-812915-9.00007-1.

30. Akhtar, J., Aamir, M., & Sher, M., (2018). Chapter 2-organometal lead halide perovskite. In: Thomas, S., & Thankappan, A., (eds.), *Perovskite Photovoltaics* (pp. 25–42). Academic Press. doi: https://doi.org/10.1016/B978-0-12-812915-9.00002-2.

31. Choi, J., Park, S., Lee, J., Hong, K., Kim, D. H., Moon, C. W., Do, P. G., et al., (2016). Organolead halide perovskites for low operating voltage multilevel resistive switching. *Adv. Mater., 28*, 6562–6567. doi: 10.1002/adma.201600859.

32. Hwang, B., & Lee, J. S., (2017). Hybrid organic-inorganic perovskite memory with long-term stability in air. *Sci. Rep., 7*, 1–7. doi: 10.1038/s41598-017-00778-5.

33. Hwang, B., & Lee, J., (2018). Lead-free, air-stable hybrid organic-inorganic perovskite resistive switching memory with ultrafast switching and multilevel data storage. *Nanoscale, 10*, 8578–8584. doi: 10.1039/c8nr00863a.

34. Xiong, Z., Hu, W., She, Y., Lin, Q., Hu, L., Tang, X., & Sun, K., (2019). Air-stable lead-free perovskite thin film based on $CsBi_3I_{10}$ and its Application in resistive switching devices. *ACS Appl. Mater. Interfaces, 11*, 30037–30044. doi: 10.1021/acsami.9b09080.

35. Choi, J., Van, L. Q., Hong, K., Moon, C. W., Han, J. S., Kwon, K. C., Cha, R. P., et al., (2017). Enhanced endurance organ lead halide perovskite resistive switching memories operable under an extremely low bending radius. *ACS Appl. Mater. Interfaces, 9*, 30764–30771. doi: 10.1021/acsami.7b08197.

36. Cai, H., Ma, G., He, Y., Liu, C., & Wang, H., (2018). A remarkable performance of CH 3 NH 3 PbI 3 perovskite memory based on passivated method. *Org. Electron. Physics, Mater. Appl., 58*, 301–305. doi: 10.1016/j.orgel.2018.04.025.

37. Landi, G., Subbiah, V., Reddy, K. S., Sorrentino, A., Sambandam, A., Ramamurthy, P. C., & Neitzert, H. C., (2018). Evidence of bipolar resistive switching memory in perovskite solar cell. *IEEE J. Electron Devices Soc., 6*, 454–463. doi: 10.1109/JEDS.2018.2820319.

38. Zhou, F., Liu, Y., Shen, X., Wang, M., Yuan, F., & Chai, Y., (2018). *Low-Voltage, Optoelectronic $CH_3 NH_3 PbI_3$–xCx Memory with Integrated Sensing and Logic Operations.* 1800080, 1–8. doi: 10.1002/adfm.201800080.

Index

A

Activation barrier, 14, 16, 94, 98, 99
Adipocytes, 45
Adjoint matrix, 136, 138
Adsorption
  isotherm, 76, 88
  kinetics, 86
Agrochemicals, 108
Aldehydic group, 79
Aliphatic hydrocarbons oxidation, 106
Alkaline conditions, 100
Alkane, 106, 111
Alkoxy radicals, 44
Alkyl group, 53
Allotropes, 180
Aluminum (Al), 122, 128, 197
Alzheimer's disease, 44
Amino groups, 72
Anisotropy, 156
Anti-Alzheimer's property, 45
Anticancer properties, 41
Anti-estrogenic activities, 43
Antioxidants, 41, 44, 45
Aromatic
  carbon atom, 53, 54
  hydrocarbons, 108
  protons, 53
  rings, 53
Arsenic solution, 75, 84, 86
Arsenicosis, 72
Atherosclerosis, 45
Atomic
  absorption spectrophotometer (AAS), 75
  orbital (AO), 28, 32
Atoms-in-molecules (AIM), 2, 29
Axial stresses, 165–168

B

Bandgap, 46, 188, 193, 194, 201
Benzene oxidation, 108, 110

Benzoic acid, 97
Bias voltage, 193, 196–199, 203
Bimolecular
  donor-acceptor reactive systems, 32
  reactive system, 14, 23
Bioactivities, 41, 42, 43, 42, 58
Biomasses, 72, 73
Bio-medical technology, 181
Boltzmann constant, 18
Bond ionicity, 28
Bones, 155–157, 168
Born-Oppenheimer (BO), 5, 14
Boundary conditions, 159, 161, 186
Bravais lattice, 182
Brillouin zone, 187
Buffer layer, 206–208

C

Carbohydrates, 44
Carbon
  atoms, 53, 54, 95, 179, 182, 186, 187
  carbon
    bond, 107–109, 180, 182, 187, 189
    interactions, 179
  chemistry, 180
  nanostructures, 179
  nanotubes (CNTs), 179–186, 189, 190
    electrochemical properties, 186
    networks, 184
Carbonyl oxygen atom, 53
Cardiovascular disease, 45
Catalysis, 93–98, 102, 105, 111–113
Catalytic
  cycle, 103, 111
  oxidation, 105
Cayley
  Hamilton-Frobenius theorem, 135
  Sylvester-Pauli matrices, 174
Central sophisticated instrumentation
  facility (CSIF), 58

C-H bond, 105
  activation, 105, 106, 108, 109
Chalcopyrite, 62, 63
Charge
  carrier density, 199
  sensitivity analysis (CSA), 3, 25
  transfer (CT), 3, 4, 24–31, 46, 72–74,
    78–81, 83–85, 87, 88
Chemical bonds, 2
Chemisorption, 89
ChemTools, 49
Cheung
  Cheung function, 120, 128
  Chung method, 121
Chirality, 187, 188
Chitosan (CT), 71, 72, 83, 89
  root powder
    composite (CTRP), 78, 79, 83, 84
    physical blend (CTRB), 78, 79, 83
Chromium oxide, 112
Circumferential stresses, 165, 167
Clusters, 93, 96, 97, 105, 106, 108, 109,
  111, 112
Communication theory, 28
  chemical bond (CTCB) , 27
Complementary metal-oxide-semicon-
  ductor (CMOS), 194, 208
Complex Riemann-Silberstein vector, 146,
  151
Conduction
  band (CB), 62, 65, 187, 188, 190
  mechanisms
    Fowler-Nordheim tunneling, 200
    ohmic conduction, 198
    Schottky thermionic emission, 199
    space charge limited conduction
      (SCLC), 195, 196, 199
    trap-filled limit region (TFL), 198
Contragradience (CG), 2
Cost-effective tools, 72
Coulombic Hamiltonian, 10
Coumastans, 41, 43
Coumestrol, 41–46, 49–58
Covalency, 28, 31
Covalent chemical bond, 17
Critical points, 159
Cylinder, 155, 156, 158, 162, 164, 168,
  181, 187

D

Dehydrogenation, 111
Dehydroisomerization, 111
Density functional theory (DFT), 2, 3, 33,
  41, 45, 48, 49, 51, 58, 61, 63–67, 71,
  73, 77, 80, 89, 99, 102, 103, 106, 109,
  181, 190
Determinicity, 7, 9, 10, 28
Dielectric constant, 199, 201
Differential equation, 147, 148
Dipole moment, 62, 63, 66
Dirac matrices, 171, 173, 174, 177
Dispersive forces, 77
Disproportionation, 30
Divacant-graphene (DVG), 103
Donor
  acceptor
    interaction, 64
    systems, 4, 23
  molecular system, 63
Dopant atom, 98
Double
  numerical plus polarization (DNP), 78,
    89
  vacancy graphene, 102
  walled CNTs, 180
Duality rotations, 141, 143, 145–147, 150,
  151
Dynamical
  equations, 6
  theory, 143

E

Effective velocity, 6
Eigenstates, 17, 18
Eigenvalue, 18, 138
Eigenvectors, 136
Elastic-plastic
  extensions, 155
  stress, 155, 156, 160
  transition phase, 168
Electrical conductivity, 103, 184, 185, 187
Electrocatalysts, 101
Electrocatalytic activity, 101
Electrochemical cell, 129

Electrode, 119, 120, 122, 127, 128, 184,
    195, 197, 198, 204, 207, 208
    active layer interface, 197
    dioxides, 198
Electromagnetic energy, 141, 143, 145,
    147, 150
Electron
    acceptor, 23, 44
    affinity (EA), 30, 47, 48, 51, 58, 63
    attracting capacity, 51
    catalytic activity, 101
    cloud, 51, 54, 108, 180, 187
    delocalization, 28
    density, 3, 7, 13, 22, 24, 26, 33, 53, 77
    donating
        capacity, 48
        power, 47, 48
    donor, 23
    hole
        diffusion, 193, 194, 201
        transport, 193
    insertion potential, 30
    localization function (ELF), 2, 97
    phonon interactions, 186
    populations, 24, 25
    transfers, 24
Electronegative atoms, 53, 54
Electronegativity, 17, 28, 48, 51, 61,
    63–66, 98
Electronic
    circuits, 184
    density, 13, 16, 95, 181
    energy, 1, 4, 10, 17, 20–22, 26, 34, 47, 181
    Hamiltonians, 5, 17, 18, 31
    spin states, 109
    states, 2–4, 8, 10, 17, 33
    structure, 22, 32, 105, 113, 189
    traps, 127
Electrophilicity, 48, 51, 62–66
Electrostatic
    interactions, 28
    potential maps, 46
Eley-Rideal (ER), 97, 99
Endurance time, 201
Ensemble-average
    energy, 17, 21, 27
    value, 3, 20

Entropic
    arguments, 14
    principles, 21, 33
    theories, 2
Entropy, 2, 3, 8–10, 12, 18, 19, 22, 23, 28,
    33
Equidensity orbital's (EO), 8, 121
Equilibria, 21, 23
Equilibrium, 3, 4, 9, 13, 17–20, 22–26, 33,
    34, 48, 51, 76, 77, 85, 86, 88, 102, 128,
    159, 161, 185, 190
    concentrations, 88
    current, 9
    data, 88
    equation, 159
    phase, 9, 13
    principles, 23, 34
Estrogenic activity, 43
Ethane, 106–108, 112
Ethene, 107, 111
Ethylene, 28, 107, 121, 129
    carbonate (EC), 121, 129
Euler-Lagrange equation, 141–146, 148
Exo-ergic
    reactions, 14, 16
    transitions, 16
Exponential energy model, 49

F

Faddeev-Sominsky
    method, 138
    techniques, 134
Faraday tensor, 150
Femur, 155–157, 164, 168
Fermi
    energy, 187, 190
    level, 95, 96
Fluorine-doped indium tin oxide (FTO),
    195, 203–205, 207, 208
Fock's space, 17, 18
Fourier transformed infrared spectropho-
    tometer (FTIR), 73, 74, 78
Fowler Nordheim tunneling, 195, 196, 200
Freundlich
    adsorption equation, 77
    models, 76, 88

Frontier
  electron (FE), 29–31, 34
  molecular orbital (FMO), 42, 46, 58, 64
Fukui
  function (FF), 17, 22, 25
  indices, 42
Fullerenes, 180
Functional
  groups, 72, 97
  moiety, 74, 89

G

Gas phase
  cluster ions, 105
  oxidation, 108
Gaussian 09 software package, 45, 57
General memory device, 195
  characteristic aspects, 195
    conduction mechanisms, 198
    memory devices structure, 195
    memory parameters, 200
    resistive switching, 196
Generalized gradient approximation
  (GGA), 77
Generic nomenclature, 182
Gibbs free energy, 105, 110
Global
  chemical potential, 22
  descriptive parameters, 47, 51, 57, 58
  descriptors, 10
  reactive
    descriptive parameters, 46, 51
    descriptors, 47
Gradient
  entropy, 9
  information content, 20
Gram-scale production, 180
Grand-ensemble, 1, 3, 17, 18, 23, 26, 34
Graphene, 94–103, 112, 113, 180–182,
  184, 186–189
  based nano material, 96
  metal system, 98
  oxide (GO), 96, 97
  sheet, 180, 181, 187
Graphite, 180–182, 187, 203
Green's function, 49, 186
Ground-state (GS), 14, 19, 22, 181

H

Hamiltonian
  equation, 187
  matrix, 186
Hammond rule, 1
Hard (soft) acids and bases (HSAB), 2, 4,
  27, 28, 32, 34
Harriman-Zumbach-Maschke (HZM), 8
Hermitian information operator, 10
Hetero-atoms, 95, 96, 100
Heterocyclic ring system, 53
Heterogeneous catalysts, 94, 96, 97, 105
Hexagonal
  network, 186
  rings, 180
High resistance state (HRS), 196–198,
  200, 204, 205, 207, 208
Highest occupied molecular orbital
  (HOMO), 29, 31, 46–48, 51, 54, 56, 57,
  61, 63–67, 127, 128
Homogeneity, 5, 7, 16
Homogeneous linear differential equation,
  147
Hybridization, 98
Hydro desulfurization, 111
Hydrocarbons, 105–108, 111, 112
Hydrogenation, 103
Hydroxylation, 108
Hydroxylic protons, 53
Hysteresis, 193, 194, 200, 205

I

Ideal memory device, 200
Ideality factor (η), 129
Ignorable variable, 144, 145, 151
Indeterminicity, 9, 28
Indium tin oxide (ITO), 119, 120,
  122–125, 127, 128, 195, 205–208
Information
  continuity, 11
  theory (IT), 2, 9, 22, 23, 32
Infrared (IR), 42, 74, 78–80, 113, 125
Inorganic
  devices, 120
  memory, 195, 205, 206
  perovskite structure, 202

Intermolecular hydrogen bonds, 83
Inter-reactant
  blocks, 32
  bonds, 32, 34
  parts, 32
Inverse matrix, 137
Ionic
  bonds, 28, 32
  stabilization, 31, 32
Ionization
  energies, 14, 51, 63
  potential (IP), 13, 14, 47, 48, 51
Irreversible thermodynamics, 12
Isoflavanones, 43
Isoflavans, 43
Isoflavones, 42, 43
Isoflavonoids, 41, 58
Isomerization, 111
Isotherm, 72, 76, 87–89

K

Kinetic energy, 1, 3, 7, 8, 10, 13, 14, 16,
  20, 33, 34
Kohn-Sham
  orbital, 13
  principle, 49
  theories, 8
Koopmans's theorem, 47, 51

L

Lagrange multipliers, 10, 20
Lanczos
  technique, 141, 143, 144, 148
  variational technique, 144, 151
Langmuir
  adsorption isotherm, 72
  Hinshelwood, 99
  isotherm, 76
Leverrier
  Faddeev's technique, 137
  Takeno's algorithm, 138
Lewis acid, 46
Light-emitting diodes (LEDs), 61, 62, 67,
  194
Linear differential equation, 141, 143, 148,
  150, 151, 156, 159

Liouville-Ostrogradski expression, 147, 148
Lithium perchlorate, 121
Local spin density approximation (LSDA),
  61, 63, 67
Lorentz transformations, 173, 175–177
Low resistance state (LRS), 196–198, 200,
  203–205, 208
Lowest unoccupied molecular orbital
  (LUMO), 29, 31, 46–48, 51, 54, 56, 57,
  61, 63–67, 127, 128

M

Macroscopic
  composite system, 26
  systems, 22, 23
Magnetic properties, 42, 61
Mark Helfrich equation, 198
Matrices, 171, 173, 174, 176, 186
Matrix, 1, 25, 26, 31, 32, 133, 134, 137,
  138, 172–174, 186
Maxwell equations, 141, 143, 145–147, 151
Melanin, 45
Melasma, 45
Memory device fabrication, 193
Mesoporosity, 181
Metachromatic dye, 121
Metal
  free catalysts, 96
  organic frameworks (MOFs), 95
  oxide, 94–96, 101, 104, 105, 107,
    111–113, 202
Methanol, 55, 57
Methylammonium (MA), 25–27, 203–205
Micrograph, 83
Minkowski space-time, 177
Modulus, 4, 6, 7, 10, 11, 13, 33
Molecular
  bonding, 127
  communication network, 28
  docking analysis, 42, 58
  dynamic (MD), 181
  electronic
    states, 1, 2, 6
    structure, 2, 9, 33
  equilibrium states, 4
  fragments, 33

geometry, 42
hardness, 29, 63, 65, 66
interactions, 2, 29
mechanics (MM), 111–113
reservoir, 24
scenarios, 22
spin-orbital's (MO), 8, 10, 28, 29, 31,
    56, 57
states, 33
systems, 2, 22, 111
weight, 45
Monovalent large cation, 201
Mucopolysaccharides, 121
Multi-scale modeling approach, 184
Multi-walled carbon nanotubes
    (MWCNTs), 180, 184
Mutual coherence, 2

N

Nanodomain, 181
Nanomanufacturing, 179
Nanomaterials, 93, 95, 112
Nanometers, 95, 180
Nano-reactor, 95
Nanoscale, 93
Nanoscopic property, 181
Nanosurface, 95
Nanotechnology, 93
Nanotubes, 179, 184
Negative
    ionization, 30
    reaction energy, 16
N-electron
    Hamiltonian, 22
    states, 8
    systems, 7, 10
Neutral cobalt oxide clusters, 108
Nitrogen
    oxides, 105
    reduction reaction (NRR), 97, 102–104
Noether
    constant, 141, 144
    theorem, 141, 143, 146–148, 150, 151
Non-equilibrium greens function (NEGF),
    185, 190
Non-Hermitian entropy operator, 10

Nonlinear
    optical devices, 61, 62
    region, 125
Non-volatile memory, 196
Norton's power law, 156
Nuclear magnetic resonance (NMR), 42,
    43, 52, 53, 58, 113
    spectroscopic analysis, 52
        C13 NMR spectral analysis, 53
        H1 NMR spectral analysis, 53
    spectroscopy, 43, 53, 58
Nuclei, 3–5, 11, 46

O

Ohmic region, 198
On-off ratio, 200, 201, 203–208
Open molecular systems, 1, 3, 17, 33, 34
Optoelectronic devices, 120
Orbital communication theory (OCT), 2, 28
Organic
    cation, 201
    compounds, 42
    diode, 119, 120, 127, 128
    dye, 127, 128
    inorganic
        hybrid perovskites, 202
        perovskite structure, 202
    materials, 120, 127
    memory, 195, 206
    molecules, 105
    semiconductors, 127, 128
Orthogonality, 8
Orthotropic
    elastic behavior, 156
    macroscopic symmetry, 155
    materials, 158, 168
    media, 158
Ortmann, Bechstedt, and Schmidt (OBS), 78
Osteoporosis, 45, 156
Oxidation reactions, 95, 97, 109, 112
Oxidative dehydrogenation, 111
Oxygen
    centered radicals, 106
    evolution reaction (OER), 97, 100–102,
        113
    insertion mechanism, 109

reduction reaction (ORR), 97, 100–102

P

Parkinson's disease, 44
Pathogenesis, 44
Percolation threshold, 184, 185
Perovskite, 193–195, 197, 198, 201, 202, 208
  active layer, 207
  materials, 193–195, 201, 208
  memory devices, 193, 201, 202, 205–208
    all inorganic perovskite memories, 205
    buffer layers role, 206
    mixed halide perovskite memory
      devices, 202
    perovskite materials, 201
    trihalide perovskite memory devices,
      204
  memory, 193–195, 197, 200–202, 205,
    207, 208
Perturbations, 3, 22, 47, 48
Petrochemicals, 108
pH, 75
Phenol, 108, 109, 112
Photodetectors, 193
Photodiode, 120
Photonics, 93
Photovoltaic, 63
  applications, 62
  properties, 62
Phytochemicals, 41
Phytoestrogens, 43
Polarization, 24, 26, 48, 78
Polyethylene oxide (PEO), 121, 129
Polymer, 72, 81, 82, 98, 120, 195, 196
Polymeric
  chains, 83
  moieties, 79
  structures, 77
  system, 82
Polymerization, 106, 111
Polyvinyl alcohol (PVA), 121, 122, 129
Pore structure, 94
Potential energy, 10, 11, 20, 49, 109, 113
  scan (PES), 15, 16, 41, 46, 50, 57, 109
  surface (PES), 16, 49, 109, 113
Principal stresses, 160, 161

Promolecular reference, 23
Propylene carbonate (PC), 121
Protonation, 103
Protons, 53, 54
Pseudo-Euclidean spaces, 138
Pseudo-first order kinetic model, 76
Pseudo-second-order kinetic models, 76,
  86, 89
Pterocarpans, 43
Pubchem software, 45

Q

Quadratic energy model, 49
Quadrupole moment, 65
Qualitative diagram, 29
Quantum
  electronic state, 187
  information theory (QIT), 2, 3, 22, 32,
    33, 34
  mechanical (QM), 1, 2, 7, 10, 11, 22, 42,
    96, 111, 112, 174, 185, 200
  state, 4, 7, 20
Quaternionic
  triple product, 171
  units, 171, 173, 174
Quaternions, 177

R

Radial stresses, 166
Radius ratio, 164–167
Rational energy model, 49
Reactants, 1, 3, 4, 23–34, 64, 78, 81, 96, 107
Reaction
  coordinate (RC), 14, 16
  partner, 23, 24
Reactive oxygen species, 44, 45
Reactivity
  criteria, 23, 34
  theory, 1, 3, 4, 14, 34
Relative
  Gibbs free energies, 110
  velocity, 176, 177
Resistance, 47, 96, 119, 121, 123, 125,
  127–129, 196, 204–206, 208
Resistive
  memory cell, 203

random access memory (RRAM), 194, 208
switching
   filamentary conduction, 197
   space charge traps, 197
Resonance energy, 108
Resultant
   gradient information, 1, 3, 4, 7, 11–13,
    15–17, 20, 22, 33, 34
   information, 7, 16, 34
Retention time, 193, 195, 200, 201, 204–206
Richardson constant, 124
Rotenoids, 43
Rund-Trautman identity, 142

S

Scalar relativistic effect, 78
Scanning electron microscope, 75
Schottky
   emission, 196, 199
   emotion, 200
Schrödinger equation (SE), 6, 11, 22, 185
Semiconductor, 62, 63, 127, 128, 188, 199
Semi-metallic character, 188
Seth's transition theory, 155, 156, 168
Silica-alumina, 95
Single-walled carbon nanotubes
   (SWCNTs), 180–184, 187–189
Solar cells, 61, 62, 67, 193, 194, 207
Solid electrolyte, 121
Sorption pattern, 72, 76
Space charge limited conduction (SCLC),
   195, 208
Spatial rotations, 176, 177
Spectroscopic tools, 71
Spintronics, 63
Stress-strain relations, 158
Structure-activity relationships, 94
Sulfur dioxide (SO_2), 105
Synergistic effect, 84, 89, 98

T

Technology computer-aided design
   (TCAD), 184, 190
Tensile strain, 207
Therapeutic applications, 41
Thermal conductivity, 95

Thermionic
   emission (TE), 121, 124, 125, 128, 129,
    195
   injection, 199
Thermo gravimetric analysis (TGA), 74,
   79, 83
Thermodynamic, 12, 21, 22, 23
   conditions, 1, 17–19, 23
   description, 23
   energy, 3
   equilibria, 3, 23, 34
   mean gradient-information, 21
   potential, 18, 19
   principles, 21
   rule, 20
   state, 3
Thermograms, 83
Thermogravimetric analyzer, 74
Thionin, 119–121, 127
Third-generation solar cells, 62
Tibia, 155–157, 164, 168
Time-dependent density functional theory
   (TDDFT), 49, 54
Toxic substances, 44
Transition
   phase, 156, 168
   state (TS), 14, 16, 99, 105, 109
Transmission electron microscopic (TEM),
   180
Trap-filled limited voltage (VTFL), 198
Trihalide perovskites, 195
Turbidity, 73, 84, 85

U

Ultraviolet (UV) spectroscopy, 43
   visible spectroscopy, 42, 54, 58
Unpaired spin density (UPSD), 106
Unsaturation, 95

V

Vacuum permittivity, 199
Valence
   band (VB), 62, 65, 187, 188
   electrons, 78
Van der Waals (vdW), 77, 181
Vanadium oxide, 106, 108, 109

Variational
  principle, 3, 10, 11, 19, 142, 144
  symmetry, 141, 143, 145, 147, 148, 151
Velocity, 6, 12–14, 145, 175
Vertical
  electron affinity (VEA), 47
  ionization potential (VIP), 47
Vibrational frequencies, 42, 78, 80
Virial theorem, 1, 4, 14, 16, 17, 33, 34
Volatile memory, 193, 196, 197

W

Water hyacinth, 73, 74, 89
  root powder (WHRP), 73, 74, 78, 79,
    83, 84, 88

Wave function, 2, 4, 6, 7, 10, 11, 13, 49
World Health Organization (WHO), 71
Write once read many (WORM), 196, 197

X

X-ray spectroscopy, 204

Y

Yield stress, 161

Z

Zeolites, 95, 108
Zig-zag CNT, 182, 183, 189
Zone folding approximation, 187

Printed in the United States
by Baker & Taylor Publisher Services